AF840110

Quantum Leaps In Biochemistry

FOUNDATIONS OF MODERN BIOCHEMISTRY
A Multi-Volume Treatise, Volume 2

Editors: **MARGERY G. ORD and LLOYD A. STOCKEN,** *Department of Biochemistry, University of Oxford, Oxford, England*

Quantum Leaps In Biochemistry

Edited by: **MARGERY G. ORD**
LLOYD A. STOCKEN
Department of Biochemistry
University of Oxford
Oxford, England

JAI PRESS INC.

Greenwich, Connecticut *London, England*

Library of Congress Cataloging-in-Publication Data

Foundations of modern biochemistry / editors, Margery G. Ord and Lloyd
A. Stocken
p. cm.
Includes bibliographical references and indexes.
ISBN 1-55938-960-5 (v.1)
1. Biochemistry—History. I. Ord, Margery G. II. Stocken, Lloyd
A.
QD415.F68 1995
574.19'2'09—dc20 95-17048
CIP

Copyright © 1996 by JAI PRESS INC.
55 Old Post Road, No. 2
Greenwich, Connecticut 06836

JAI PRESS LTD.
The Courtyard
29 High Street
Hampton Hill, Middlesex TW12 1PD
England

All rights reserved. No part of this publication may be reproduced, stored on a retrieval system, or transmitted in any form or by any means, electronic, mechanical, photocopying, filming, recording or otherwise without prior permission in writinq from the publisher.

ISBN: 0-7623-0077-9

Transferred to digital printing 2006

CONTENTS

LIST OF CONTRIBUTORS

Iain D. Campbell	Department of Biochemistry University of Oxford Oxford, England
Anil Day	School of Biological Sciences The University of Manchester Manchester, England
R. A. Dwek	Department of Biochemistry University of Oxford Oxford, England
J. Murdoch Mitchison	Institute of Cell, Animal and Population Biology University of Edinburgh Edinburgh, Scotland
Margery G. Ord	Department of Biochemistry University of Oxford Oxford, England
Joanna Poulton	Department of Pediatrics The John Radcliffe Hospital Oxford, England
Philip Siekevitz	The Rockefeller University New York, New York
Lloyd A. Stocken	Department of Biochemistry University of Oxford Oxford, England
Jan A. Witkowski	The Banbury Center Cold Spring Harbor Laboratory Cold Spring Harbor, New York

ACKNOWLEDGMENTS*

Once again we thank Professor Radda for continuing to allow us space in the department and our colleagues whom we consulted in the preparation of this volume, especially Professors Sir Henry Harris and Ed Southern and Dr. Michael Yudkin. On a recent visit to Oxford Dr. J. D. Watson also gave us helpful advice.

We are very grateful to Drs. Bruce Henning, Cathy Pears, and Michael Yudkin for reading parts of our manuscript, to Ms A. Morgan for her photography, Mr. Brian Taylor and the staff of the Radcliffe Science Library for very patient help over references, and the members of the University Computing Service for assistance.

Dr. Siekevitz compiled the references to protein synthesis and ribosomes for the chronology; he and the Rockefeller Archives Center kindly provided photographs of Max Bergmann, Albert Claude, Joseph Fruton, George Palade, Keith Porter and Philip Seikevitz. Dr. Witkowski and the Cold Spring Harbor Laboratory gave us the pictures of Stanley Cohen, Walter Gilbert, Tom Maniatis, Kary Mullis, Don Nathans, Rich Roberts, Ed Southern, and Howard Temin. Somerville College, Oxford, provided the picture of Dorothy Hodgkin; Dr. Anil Day obtained the photo of Boris Ephrussi; Dr. Szpirer assisted us in obtaining a picture of Jean Brachet; and Drs. Kornberg and Zamecnik sent us their photographs. We are also grateful to the Nobel Foundation for permission to reproduce their photographs of Francis Crick, Alfred Hershey, Robert Holley, John Kendrew, Gobind Khorana, Marshall Nirenberg, Severo Ochoa, Max Perutz, E. L. Tatum, and Jim Watson. The photograph of Rosalind Franklin was reproduced from J. F. Judson's "The 8th Day of Creation". Those of George Beadle, Erwin Chargaff, Arthur Kornberg, and Fred Sanger are reproduced, with permission, from Volumes 43 (1974), 44 (1975), 57 (1988), and 58 (1989) of the *Annual Reviews of Biochemistry*. The photograph of Dan Mazia by Paul Maurer was obtained through the good offices of Dan Mazia and Murdoch Mitchison.

Margery G. Ord
Lloyd A. Stocken
Editors

*Superscript numbers next to surnames throughout this volume refer to photographs, pages 99–107.

Chapter 1

INTRODUCTION

In Volume 1, *Early Adventures in Biochemistry*, we described the experimental methods used in the elucidation of the main pathways of intermediary metabolism in animals. We drew attention to the "forgotten men of biochemistry" and their achievements, and tried to show younger biochemists how, in spite of very primitive equipment, certain fundamental concepts were advanced. These were that ATP was the primary energy source for chemical and physical work done by cells, that proteins were the workhorses of the cell and contributed significantly to the structures from which the cells are composed, and that events in cells were spatially and temporally organized.

Three further propositions that emerged during the 1950s were only touched on in the first volume—the role of DNA as the carrier of the inherited information of the cell, the metabolic activity of the different species of RNA, particularly the role of the ribosome in protein synthesis, and the ideas of Jacob and Monod regarding the regulation of expression from the genome. The first two of these are now considered in more detail in Chapters 2, 3, and 5 of this volume.

The exponential growth of molecular biology followed from the development of experimental techniques for analyzing the nucleotide sequences of DNA. The various procedures by which "foreign" DNA can be introduced into and expressed by host cells are reviewed in Chapter 3. We hope to consider regulation of expression of the genome in Volume 3.

The finding of extracellular DNAs in mitochondria and chloroplasts (Chapter 4), and the establishment of their probable endosymbiotic origins, was followed by the discovery, in plants and lower organisms, of the movement of DNA molecules between plastids, mitochondria, and nuclei. That mechanisms exist for interchange between plastid and nuclear DNA was another discovery of great evolutionary significance.

Other important experimental innovations include the application of nuclear magnetic resonance (NMR) to the study of protein structure. NMR has provided what is currently the most powerful method for examining protein interactions with macromolecules, substrates, and other solutes. The results of such studies, together with our present ability to deduce protein sequences and likely functions from genetic data, have led to a major change in our thinking about proteins. Up to 1960 attention was primarily focused on the properties of enzymes and their mechanisms of action. In the past 20 years many more proteins have been discovered, of which most occur only in small amounts in cells and probably have regulatory roles either in the nucleus or in processes whereby extracellular events at the cell surface lead to intracellular responses. Most of these proteins are not enzymes: instead their effects are exerted through contacts with other proteins or cell constituents. Genetic and structural analyses, aided by highly sophisticated computer techniques, now concentrate on protein domains (see Doolittle, 1995). The size and shape of these domains makes them analyzable by NMR (Chapter 6), which, along with X-ray crystallography, has been an important means by which these regions have been identified.

Glycobiology is a striking example of a branch of biochemical research whose existence has been almost totally dependent on the introduction of novel analytical methods (see Chapter 7). In the 1970s glycosylated molecules, usually complex mixtures of closely related compounds, were difficult to separate and whose precise composition defied analysis. These problems are now largely overcome, determinants for protein glycosylation are emerging, and its tissue and species diversity at different stages of normal or pathological development can now be examined.

The integration of the synthesis of proteins and their migration to the appropriate regions of the cell, or for export, is considered, *inter alia*, in Chapter 5. A further topic in cell biology—the way in which the behavior of the cell is directed successively towards growth, DNA replication, and cell division—is discussed in Chapter 8. Analysis of the cell cycle illustrates the way in which advances in biochemistry have utilized the full range of classical, genetic, and physical methods.

The first volume drew attention to the work of early biochemists who established metabolic pathways using very simple apparatus. This volume covers some of the phenomenal advances made since the 1950s, facilitated in large part by the expansion in the 1960s both in numbers of scientists and in available resources.

Since many of the above areas of research are still under active investigation, we have asked the contributors to focus on what appear to them to be the conceptually significant developments and how these were achieved, and not to attempt an up-to-the-minute coverage of each topic. Their long-term experience has produced authoritative accounts of the quantum leaps made in their fields.

REFERENCES

Doolittle, R.F. (1995). The multiplicity of domains in proteins. Annu. Rev. Biochem. 64, 287–314.

Chapter 2

THE CODING PROPERTIES OF DNA AND THE CENTRAL DOGMA

Margery G. Ord and Lloyd A. Stocken

INTRODUCTION

This chapter is concerned with observations prior to 1953 which indicated a role for DNA in information transfer, and the experiments (up to 1980) which validated the Watson and Crick structure for DNA and its consequences.

INFORMATION STORAGE AND TRANSFER BEFORE 1953

Nuclei, first isolated by Miescher in 1869, were found to contain a phosphorus-rich substance, nuclein. When similar material was analyzed from salmon sperm, two components were distinguished—an acidic phosphorus-containing nucleic acid and a basic protein, protamine. Thymonucleic acid from thymus glands contained phosphorus; the bases thymine, cytosine, adenine, and guanine; and the pentose sugar, 2-deoxyribose-DNA. The nucleic acid obtained from yeast, RNA, contained uracil, not thymine, and ribose rather than deoxyribose.

That DNA and protein were the major components of chromosomes became evident from cytochemical staining and UV microscopy in the 1920s and 1930s. The preparation of nucleic acids, free from traces of protein, was however extremely difficult. Both DNA and especially RNA were easily degraded during isolation, and methods for their analysis were extremely primitive. Determinations of the nitrogen and phosphorus contents of DNA were consistent with a nucleotide structure, and analyses of the bases indicated roughly equimolar proportions of purines and pyrimidines. By the 1930s a tetranucleotide structure for DNA had therefore been proposed by Levene. Since this did not appear to allow the range of protein diversity already apparent, it was supposed that inherited information was a property of the protein(s) of the chromosomes, not of the DNA (For refs., see Ord and Stocken, 1995).

The experiments of Griffiths (1928) on mice infected with pneumococci showed that information could be transferred between cells. Small numbers of living pneumococci type II (rough coated), which did not cause fatal bacteremia, were injected into mice together with a large inoculum of heat-inactivated (killed) type III (smooth coated) pneumococci. Blood from animals which subsequently died yielded pure cultures of type III, virulent, bacteria. Later experiments showed that cell-free extracts from the virulent strain could carry out the transformation. In 1944, Avery, McLeod, and McCarty established that extracts which had been virtually freed from protein by chloroform, and which contained neither detectable lipid nor serologically identifiable polysaccharide, brought about transformation. The transforming principle was resistant to hydrolysis by RNAase, trypsin, or chymotrypsin, but was destroyed by DNAase, i.e. it appeared to be DNA. Once transformed, the pneumococci could be propagated as the smooth, encapsulated strain without further exposure to the transforming principle.

In spite of this apparently clear-cut demonstration of the capacity of DNA to transform cells, the possible presence of small amounts of protein in the extract could not be excluded. With the limited knowledge of its structure then available, those who were unable to accept that DNA could carry the necessary information to cause transformation were still able to attribute the change to protein in the extract.

Explicit evidence for the ability of DNA to transform came from the neat experiments of Hershey[6] and Chase (1952) using T_{even} bacteriophage grown in $[^{32}P]P_i$ to label the DNA and ^{35}S-methionine to label the protein of the viral coat. The radioactive phage was then harvested and used to infect unlabeled *E. coli*. All the ^{32}P-labeled DNA entered the bacterium, but the ^{35}S-protein coat of the virus adhered to the outside of the cell and could be shaken off by agitation in a Waring blender. No labeled sulfur was detected in the new protein of the viral particle, which must therefore have been programmed by the entering DNA.

Amounts of DNA/cell showed that nuclei from different organisms contained different amounts of DNA/nucleus, and that in a given species the amount of DNA/diploid cell was twice that in a haploid.

There were also indications of a role for RNA in protein synthesis—the presence of DNA was not essential. In 1934, in experiments with *Acetobularia*, a photosynthetic marine organism, Hammerling showed that, provided light was available, if the rhizoid containing the nucleus was removed, the remaining stalk was able to elongate (grow) and differentiate with a mushroom-like cap. The enucleated organism was however incapable of sexual reproduction, i.e. it could not sporulate (see Hammerling, 1953). Similar experiments were performed with *Amoeba*. Here, enucleated portions were still capable of some protein synthesis. Survival times though, were much shorter than with *Acetobularia* as enucleated *Amoeba* cannot feed.

By 1941, Caspersson using UV microscopy and Brachet with cytochemical staining had demonstrated RNA was present both in the nucleolus and the cytoplasm (see Caspersson, 1950; Brachet,[2] 1957). Cells with a high capacity to synthesize protein, like the parenchymal cells of the liver and pancreas, contained relatively large amounts of RNA.

One further link between nucleic acids and protein synthesis was suggested from the work of Beadle[1] and Tatum[3] (1941) (see Beadle, 1945) on X-ray or UV-induced mutants of the bread mold, *Neurospora*. The haploid spores were irradiated, plated onto a complete synthetic medium to promote growth, and then replated onto a minimal medium. At least 100 different mutants were isolated with lesions in their ability to synthesize amino acids, vitamins, or purine or pyrimidine bases, which therefore had to be added to the minimal medium to permit growth. Beadle and Tatum concluded there was a one-to-one relation between a gene and a specific reaction in the cell—one gene, one enzyme.

THE STRUCTURE OF DNA: ITS VERIFICATION AND IMPLICATIONS

A very full account of the events leading to the Watson[7] and Crick[5] hypothesis for the structure and role of DNA and its validation is given in *The Eighth Day of Creation* (Judson, 1979). Judson also stresses the vital contribution of physicists and geneticists to the story, complementing that of more traditional biochemists.

To understand how DNA carried the information for transformation, it was imperative to determine its structure. Even the degraded specimens of DNA then available had molecular weights of ca. 1×10^6 kDa, more than an order of magnitude larger than those of the proteins whose primary structures were becoming known through Sanger's sequencing techniques (Sanger,[12] 1952). Moreover the nucleases then known had very limited specificities; they could not be used to generate overlapping families of polynucleotides similar to the peptides obtained in the protein field. X-ray crystallography was therefore the only means to gain insight to the structure of DNA. This technique, however, could not indicate the order of the individual bases.

Getting good, reproducible fiber preparations proved difficult. Early pictures, such as those available to Pauling and Corey (1953), provided inadequate resolution. Better diffraction patterns were obtained by the groups from Kings' College, London (Franklin[8] and Gosling, 1953; Wilkins et al., 1953; see Watson and Crick, 1953). The patterns obtained by Rosalind Franklin for the more hydrated B form were made available to Watson and Crick (see Sayre, 1975). These, and stereochemical considerations supported by model building, led them to propose the double helical structure for DNA. They placed the bases inside and the phosphate groups outside to minimize repulsion (contrast Pauling and Corey, 1953), with the two chains running in opposite directions. They also followed the suggestion of Donohue (see Judson, 1979) that the bases should be in their keto rather than their enol form. A base from one chain would be H-bonded to a base from the other chain. "If . . . the bases only occur in the structure in the most plausible tautomeric forms . . . only specific pairs of bases can bond together . . . adenine with thymine and guanine with cytosine".

Such an arrangement was consistent with chemical analyses of DNA from several different sources by Chargaff[4] (1949–1950) and Wyatt (1952), which showed the amount of adenine equalled that of thymidine, and of guanine equalled cytosine (see Chargaff and Davidson, 1955). Chargaff indeed commented in 1950, ". . the question will become pertinent . . . whether it [A/T and G/C = 1] is an expression of certain structural principles." Watson and Crick also observed, "It has not escaped our attention that the specific pairing we have postulated immediately suggests a possible copying mechanism for the genetic material" (see Watson, 1968).

It was this latter prediction, implying semi-conservative replication, which was the first to be tested. Taylor et al. (1957) grew *Vicia faba* seedlings in a medium containing ^{3}HTdR. After thorough washing, the seedlings were transferred to medium with unlabeled thymidine and colchicine. Colchicine inhibits spindle fiber formation and thus the anaphase separation of sister chromatids. After 10 h, autoradiography showed ^{3}H-activity was equally distributed between the two daughter chromatids at the first metaphase. After 34 h, the grains were located over one only of each pair of daughter chromatids, as would be expected if the strands of the helix separated to become the templates for the synthesis of the new strands of DNA.

The following year Meselson and Stahl (1958) studied DNA replication in *E. coli* using $^{15}NH_4Cl$ as the sole nitrogen source. After growth the cells were transferred to a ^{14}N-medium, and the DNA isolated and sedimented through CsCl gradients, which allowed ^{15}N- and ^{14}N-labeled DNA to be distinguished. After one generation, 50% of the DNA had banded in the ^{15}N-position, and after two generations, the amounts of unlabeled ($^{14}N/^{14}N$) and half-labeled ($^{15}N/^{14}N$) were equal, as predicted by semiconservative replication.

DNA replication requires an enzyme system for its operation. The first DNA polymerase was isolated by Arthur Kornberg[9] in 1958 (see Kornberg, 1968). The enzyme, from *E. coli*, catalyzed the incorporation of deoxynucleoside phosphate into DNA, in an order determined by an obligatory DNA template. Polymerase activity was increased in the presence of all four deoxynucleoside triphosphates and ATP. Nearest-neighbor analysis showed that the pattern of nucleotide incorporation was complementary to that in the template strand. In this procedure, ^{32}P α-labeled deoxynucleoside triphosphates were used in turn. The product was digested with micrococcal nuclease and spleen diesterase to yield 3′-deoxynucleotides which could be separated by paper electrophoresis. ^{32}P was found attached to the 3′ neighbor adjacent to the entering nucleotide. All 16 possible arrangements of deoxynucleotides were detected, i.e. there were no forbidden sequences, and the sense of the strands showed them to be anti-parallel. The Kornberg enzyme did not however satisfy all the requirements for DNA synthesis *in vivo*; later work showed the need for other polymerases (see below).

THE DISCOVERY OF THE CODE

The publication of *A Structure for DNA* was the start of a revolution in scientific thought, taking off fairly slowly but gaining momentum from the late 1950s. "How did DNA code for amino acids?" was an immediate intellectual challenge, initially eliciting hypothetical solutions from those with cryptographic inclinations but no biochemical training. From the start it was accepted that programs would only be required for 20 amino acids. Derivatives like phosphoserine or hydroxyproline were assumed, correctly, to be formed post-translationally when the amino acids were already incorporated into the protein.

Gamow (1954) was one of the first to suggest a system of codes based on specific, distinguishable steric interactions between amino acids and DNA. Such a code would be overlapping,

A B C D A B C

and degenerate. One amino acid would be specified by more than one codon.

As more protein sequences emerged, especially the variant forms of hemoglobin with point mutations studied by Ingram, it became evident that overlapping codons were out. For example in sickle cell anemia only a change in a single amino acid was detected, i.e. glutamate 6 was substituted by valine. Point mutations were also found by Wittman in an extensive study of tobacco mosaic virus (TMV) mutants produced by nitrous acid, which deaminates and converts adenine to inosine, which mimics guanine, and changes cytosine to uracil. Further, no evidence for restrictions on amino acid neighbors was apparent, though not all possible partners were equally common (see Crick, 1963).

In 1958, Crick et al. offered a "comma-free" code. Since the four bases could, if used as triplets, code for 64 amino acids, they assumed that for any selection of three bases, only one combination from ABC, BCA and CBA was allowed. Further forbidden combinations were AAA, BBB, CCC, DDD. If the codons were triplet, this would provide nonoverlapping, nondegenerate codons for 20 amino acids. Any mutations would lead to nonsense.

Analysis of tobacco necrosis satellite virus strongly supported a triplet code. RNA from this virus contains 1200 nucleotides; it codes for a coat protein with a chain length of 400 amino acids. Further suggestive evidence that codons were triplet came from acridine-induced mutants in the rII locus of bacteriophage T4 (Crick, 1963). Acridines intercalate into DNA and cause frame-shift mutants arising from base insertions (+) or deletions (–). Changes to single nucleotides caused the synthesis of defective viral coats; a second mutation in the opposite sense (– or +) allowed intragenic suppression of the mutant. The resulting viral plaques had different appearances from normal on *E. coli* plates and were pseudo wild-type. With three mutations in the same sense in the same gene, the correct reading frame was restored, conforming with a triplet or $(3)_n$ codon.

By this time some other very important experimental developments had occurred. Gierer and Schramm (1956) succeeded in reconstructing TMV from its constituent coat protein and RNA. If the virus was reconstructed with RNA from a second strain, the proteins of the new viral particles were those of the donor RNA strain (Fraenkel-Conrat and Singer, 1957), i.e. RNA could program protein synthesis. Although it was at first thought that amino acids would interact with DNA, or more probably with RNA which seemed to be directly involved in protein synthesis, there was no convincing evidence for this. Crick (1958) therefore suggested "[an] amino acid is carried to the template by [its] adaptor molecule, and that this adaptor is the part which actually fits on the RNA." If the adaptor itself was RNA it could join onto the template by base pairing.

Isotopic evidence had shown protein synthesis to occur on ribosomes, which could be obtained after differential centrifugation of microsomes in 0.25 M sucrose medium at 10^5 g (see Siekevitz[27] and Palade,[31] 1960).

In confirmation of an earlier suggestion from Lipmann (1941) that amino acids required activation by ATP before incorporation, Zamecnik[28] and colleagues isolated an activating enzyme system which was precipitated at pH 5.0 from the postmicrosomal supernatant (see Chapter 5). This fraction contained both low molecular weight RNAs (soluble, now transfer—tRNAs) and the enzymes necessary to transfer the amino acids to these adaptors. Cell-free protein synthesizing systems were thus obtained from *E. coli* and reticulocytes.

Protein-synthesizing systems from *E. coli* were more easily purified than those from reticulocytes. DNA could be removed with DNAase and the ribosomes then sedimented and washed. Washing removed almost all the lower molecular weight endogenous RNA bound on the ribosomes (mRNA)—something which was much more difficult to achieve with reticulocytes.

Very careful analysis of the system (Matthei and Nirenberg,[15] 1961) showed that amino acid incorporation into trichloroacetic acid (TCA)-precipitable material was prevented if the preparation was treated with RNAase. It was also inhibited by puromycin and chloramphenicol, which had by then been shown to block protein synthesis in *E. coli*. A natural RNA, such as yeast ribosomal RNA (rRNA), stimulated ^{14}C-valine uptake. When synthetic polyuridylic acid (poly U) was used, ^{14}C-phenylalanine was preferentially incorporated into a product containing peptide bonds, proving that poly U selectively directed the incorporation of phenylalanine into protein.

Enzymic synthesis of polyribonucleotides became possible following the isolation by Grunberg-Manago and Ochoa[11] in 1955, of a microbial enzyme, polynucleotide phosphorylase, which catalyzed a reversible reaction:

$$n(\text{X-R-P-P}) \Leftrightarrow (\text{X-R-P})_n + n\text{P}_i$$

(see Grunberg-Manago, 1963). While it is probable that the enzyme catalyzes the phosphorolysis of polyribonucleotides *in vivo*, it could be used *in vitro* to synthesize polynucleotides, either homopolymers or, if the reaction was started with a mixture of riboside diphosphates, a heteropolymer was formed whose composition predominantly reflected that of the input mixture. The precise arrangement of the bases within such polyribonucleotides was not known.

These synthetic polynucleotides were therefore used by Nirenberg's group and by Ochoa and his colleagues (Lengyel et al., 1961) in the *E. coli* system. One out of the mixture of 20 amino acids was radioactively labeled in turn to determine which corresponded to the polynucleotide being tested. It was possible to calculate the probable composition of the polynucleotide from the ratio of the input XDPs. Triplet codons containing U were suggested for 19/20 of the amino acids (see Appendix 2 for list of codons). The commoner amino acids like alanine, glycine and serine, responded to more than one codon, demonstrating that the code was degenerate.

Unfortunately there were serious limitations to the system. Polynucleotides rich in G were difficult to prepare and those which were cytosine-rich were prone to form secondary structures, thus were much less useful. Unambiguous assignments were therefore slanted to A/U codons. Further, the procedure for precipitating protein at the end of the reaction did not allow low molecular weight di- or tripeptides to be recovered. Incomplete chains might therefore be missed. Also, the experiments were performed at relatively high Mg^{2+} concentrations, $\geq$10 mM. Under these conditions normal initiation of protein synthesis was bypassed. The need for a "start" codon was not therefore detected. In spite of all this, the results were tremendously exciting.

By 1959 Khorana[13] and his associates had developed a procedure for the unequivocal synthesis of polyribonucleotides *in vitro* (Khorana, 1959). The reactive 2′ OH groups were masked to prevent 2′→5′ joining, and condensation between units was promoted by carbodiimides, after which the masking groups were dissociated. Trinucleotides of defined order were then used by Nirenberg and Leder (1964) in a new, triplet binding, assay.

For this, ribosomes were incubated with the mixture of amino acids as before, together with the triplet under test. At the end of the reaction ribosomes carrying bound ^{14}C-amino-acyl tRNA were separated from free tRNAs by filtration using cellulose nitrate membranes. Trinucleotides were the smallest polyribonucleotides to cause unequivocal attachment of specific amino acyl tRNAs. In this way triplet codes were identified for all 20 amino acids and the direction of the reading frame determined—pGpUpU caused val tRNA binding but pUpUpG did not. The code was read 5′→3′. The codons were unambiguous and codon degeneracy was unequivocally established. Some rules for degeneracy emerged (see Woese, 1967); amino acids coded by XYU were also coded by XYC, and sometimes XYA ≡ XYG. Methionine and tryptophan had only one codon and UAA, UAG, and UGA were terminator, stop, codons.

The assignments were supported by further experiments from Khorana's laboratory, where alternating bases such as ACACA . . . were tested in the protein synthesizing system. The amino acids threonine and histidine, coded by ACA and CAC, were incorporated into an alternating peptide thr.his.thr.his . . . .

Three further properties of the coding system were defined (Woese, 1967). First, the code appeared to be universal. Various RNAs of viral origin were translated by *E. coli* ribosomes, and leucyl tRNA from *E. coli* was utilized in the rabbit reticulocyte system. Only after mitochondrial DNA had been recognized and mitochondrial protein synthesis studied (see Chapter 4) did it emerge that exceptions to codon universality occasionally occurred—e.g., the stop codon UGA is read as tryptophan in mitochondria. Exceptions have now also been found in some *Ciliophora*.

Second, the code was colinear. The order of the codons corresponded to that of the amino acids in the protein. This was implicit from Crick and colleagues' experiments with acridine mutants and was proved by Yanofsky's group in their study of mutants in tryptophan synthase in *E. coli* (see Yanofsky, 1967). The 267 amino acid residues of the A-chain of the synthase were sequenced. Mutants affecting this chain were analyzed by genetic recombination and their arrangement compared to the sites of amino acid replacement in the mutant proteins. The orders coincided.

Third, the code is read from a fixed point—something which became clearer when lower Mg^{2+} concentrations (<10 mM) were used in the protein synthesizing system so that 70s ribosomes dissociated into the 30 and 50s subunits. By 1963, Waller had observed that many of the proteins of *E. coli* had methionine as their N-terminal end. Also, when the coat protein of an RNA phage was synthesized *in vitro*, the chain began with met-ala-ser, but when the protein was isolated from the virus particles it began ala-ser. Webster and colleagues then reported that the N-terminal methionine was protected by formylation, thus ensuring that the incoming amino acid condensed only with the carboxyl group of the methionine. In the same year Marker et al. (1966) found two different tRNAs for methionine in *E. coli*, both recognizing the methionine codon AUG, but only one of which allowed the N-terminal methionine to be formylated. N-formyl $tRNA_f$ markedly stimulated amino acid incorporation into protein at low $[Mg^{2+}]$. The formyl group and often the methionine are subsequently removed to yield mature proteins. Fuller details of the prokaryotic system and the rather different one in eukaryotes are described in Chapter 5 (see also Marker et al., 1966).

One further attribute of the coding properties of tRNA will be mentioned. As structures of tRNA were determined (yeast ala tRNA, 1965; see Holley,[10] 1968) it became clear that the same tRNAs might recognize more than one codon. Yeast ala tRNA binds to GCU, GCC, and GCA. Crick (1966) suggested pairing requirements for position 3 of the codon might be less stringent than for positions 1 and 2. In ala tRNA, inosine in the first position of the anticodon, IGC, can bind to U, C, or A in position 3 of the messenger codon. By this "Wobble" hypothesis also, uracil can bind to A and G and guanine to U and C. Anticodons with U and G in their first positions can thus each recognize two codons in the third position of mRNA.

THE CENTRAL DOGMA

The central dogma, as it came to be called, was first explicitly stated by Crick in 1958: "The transfer of information from nucleic acid to nucleic acid and from nucleic acid to protein may be possible, but not that from protein to nucleic acid nor from protein to protein." In 1958, direct transfer from DNA to protein was still being considered by Gamow (see above). Also, in contrast to the results with *Acetobularia*, in thymus nuclei Mirsky and his colleagues reported amino acid uptake into protein appeared to be dependent on the presence of DNA. This was later shown to be a consequence of the manner by which these nuclei obtained their ATP. Between 1958 and 1965 adaptor (transfer) tRNA and messenger mRNA were discovered. The central dogma therefore came to be formulated: DNA → RNA → protein. Information flowed from DNA to RNA and all protein sequences were determined by RNA templates (see Watson, 1965).

The discovery of tRNA has already been mentioned. The existence of mRNA was postulated by Jacob and Monod in 1961 in their classical paper on the control of genetic expression. Until that time each gene was thought to control the synthesis of one kind of specialized ribosome, which in turn directed the synthesis of the corresponding protein. Such an idea conflicted with analytical data showing the stability and homogeneity in composition and size of ribosomal, rRNA, apparently unrelated to the composition of DNA in various organisms (Belozerskii and Spirin, 1958; see Judson, 1979). It also conflicted with views concerning the regulation of protein synthesis at the level of an informational intermediate, not at the level of protein.

There were already indications that unstable nonprotein molecules participated in the synthesis of inducible proteins in *E. coli* (see Pardee, 1985). That these molecules might be nucleic acid could be inferred from Pardee's study of pyrimidine-less mutants in *E. coli*, which required exogenous pyrimidines for adaptive enzyme formation. Volkin and Astrachan (1956) showed more directly that when *E. coli* was infected with T2 bacteriophage, there was an immediate obligatory synthesis of new RNA whose base composition resembled that of the infecting phage DNA, not that of the host. This was confirmed by later experiments with T4, where the same frequency of dinucleotide pairs was found in the new RNA as in the infecting phage DNA. T2 and T4 were again used by Brenner et al. (1961). *E. coli* were grown in a medium containing ^{15}N, ^{13}C, and ^{32}P, and, after infection, immediately transferred to "light" medium. The ribosomes were extracted in 10 mM Mg^{2+} and analyzed on CsCl gradients. After sedimentation for 35 h at 37,000 rpm, "light" and "heavy" ribosomes were separated. The bulk of the new RNA was associated with the lighter ribosome fraction. When this was dialyzed in lower 0.5 mM Mg^{2+}, the ribosomal subunits dissociated and newly synthesized RNA with MW ≡ 12s, separated. Its base composition corresponded to that of the phage DNA. The new mRNA was more sensitive to RNAase than ribosomal RNA.

mRNA in Eukaryotic Systems

The need for RNA to mediate between nuclear DNA and the sites of protein synthesis on the ribosomes was immediately apparent for eukaryotic systems, but was harder to detect than with microorganisms because of the more persistent presence of bound endogenous mRNAs. The enucleate reticulocyte, which almost exclusively synthesizes hemoglobin and is relatively free from RNAases, was the obvious system to choose. In 1961, Dintzis used very short pulses of labeled amino acids with rabbit reticulocyte ribosomes and found incorporation into the hemoglobin was almost exclusively into the carboxyl end of the molecule. He thus concluded that peptide chain growth proceeded steadily from the amino terminal end of the molecule at a rate he calculated to be about 2 residues/s.

When reticulocyte RNA was separated on sucrose gradients, RNA sedimenting at <18s stimulated amino acid uptake (see Chantrenne et al., 1967). Anemic rabbits with an enhanced reticulocyte count were next used. At 10–20 h before bleeding they were injected with $^{32}P_i$. When the reticulocyte RNA was separated on the gradients, all the RNA was labeled with the peak in radioactivity sedimenting between 4 and 16s, and with a size appropriate for coding the α- and β-globin chains. Unfortunately insufficient material was recovered to demonstrate unequivocally that the protein synthesized was rabbit hemoglobin.

Very elegant experiments by Gurdon and his colleagues (Lane et al., 1971) conclusively demonstrated that RNA from the reticulocyte system, sedimenting at 9s, actually coded for rabbit Hb. Living oocytes from *Xenopus laevis* were injected with reticulocyte RNA—4–5s; 9, 18, and 28s; or polyribosomes. The eggs were also injected with ^{3}H-histidine. After 10 h only 9s RNA and the polysomes caused marked incorporation into hemoglobin-like material. The product was examined by acrylamide gel electrophoresis and chromatography on carboxymethyl cellulose, which separate rabbit Hb from that of the frog. ^{3}H-activity was convincingly located in the region to which marker rabbit hemoglobin migrated. By 1970 evidence for mRNA had been found in many eukaryotic systems.

According to the Central Dogma, mRNA must have been transcribed from DNA. As early as 1949, Marshak and Calvet, and Barnum and Huseby observed much higher ^{32}P-turnover in nuclear than cytoplasmic RNA. Various experimental approaches indicated the nucleolus was the site of rRNA synthesis (for refs., see Abrams, 1961). Fitzgerald and Vinijchaikul used autoradiography to show ^{3}H-cytidine uptake by pancreatic acinar cell nucleoli preceded the appearance of the label in the cytoplasm. When HeLa cells were subjected to microbeam UV irradiation, irradiating the nucleoli caused a 30% fall in cytidine uptake in nuclear RNA and a 65% drop in incorporation into cytoplasmic RNA. Exposing an equal area outside the nucleolus had no such effect (see Perry, 1969). Homozygous anucleolate mutants of *Xenopus* had no nucleolar organizers and were unable to make rRNA. The tadpoles did not develop beyond the tail-bud stage (Brown and Gurdon, 1964). Similar results were found with *Drosophila* mutants with deleted nucleolar organizers.

Detailed examination of nuclei showed they contained other families of RNA besides nucleolar RNA. When ^{3}H-uridine was used to label HeLa cells, and the RNA from different sites separated, incorporation was seen first (5 min) in nucleoplasmic RNA. By 15 min there was a major peak at 45s in nucleolar RNA and in the 4s (tRNA) of cytoplasmic RNA (see Darnell, 1976). It was already clearly established that ribosomal RNA was synthesized in nucleoli. Radioactivity in nucleoplasmic, non-nucleolar RNA sedimented in a broad peak from 4 → 70s. It was therefore called heterogeneous nuclear RNA (HnRNA). Between 1959 and 1962 Harris and Watts (see Harris, 1968) examined the kinetics of isotope uptake into HnRNA and showed that much of this was degraded in the nucleus with only a small part passing into the cytoplasm.

Various laboratories reported eukaryotic mRNA was rich in adenylic acid and showed that adenyl incorporation from ATP into nuclear RNA was rather resistant to RNAase attack. The adenyl residues were thought to be at the 3′ ends of the RNA chains (see Brawerman, 1974). Studies on polyadenylated RNA were facilitated by its selective retention on poly T cellulose (Edmonds and Caramela, 1969). mRNAs from eukaryotes were also shown to have modified 5′ termini, with a methylated guanine cap, m^7GpppN′(m)-N″(m). The significance of these modifications and the mechanism by which HnRNA is processed in the nucleus, using a further class of small, stable, nuclear RNAs (snRNAs), are outside the scope of this chapter.

As originally postulated by Jacob and Monod for bacterial systems, mRNA molecules would be short-lived. In eukaryotes, mRNAs were evidently of variable half-lives. Those of *Acetobularia* and reticulocytes must be long-lasting, but others, especially some of those programming proteins involved transiently in the cell cycle (see Chapter 8), have half-lives of only a few minutes.

It was also an obvious requirement for mRNA that its sequence was complementary to that in DNA. Before sequencing techniques for DNA became available, this was most conveniently shown by hybridization.

Nucleic Acid Hybridization

By 1960 there had been considerable progress in procedures for isolating undegraded, minimally sheared DNA. Kirby used deproteinization by buffered phenol to prepare DNA from animal and viral sources. Marmur employed chloroform/isoamyl alcohol to obtain protein-free DNA from microorganisms. Viral and microbial DNA sedimented as single bands on a CsCl gradient. The buoyant density of the DNA was directly correlated with its GC content. DNA from eukaryotes was heterodisperse; sometimes distinct peaks ("satellites") separated from the bulk of the DNA, notably a band from mouse DNA (see Kit, 1963) and a crab DNA satellite which was 97% AT. These sequences occur in blocks of about 100 residues and may be repeated 10^7 times per cell.

Nucleic acids strongly absorb ultraviolet light at 260 nm. If absorbtion by DNA is followed as the temperature is increased, the absorbtion increases, reaching a plateau at a value c.35% greater than that at room temperature. This hyperchromic effect, which is seen in undegraded specimens of DNA, arises because UV absorbtion by the purine and pyrimidine bases is constrained in the double helical structure. If the DNA duplex is dissociated by, for example, heat, the constraints are removed and the UV absorbtion rises to that predicted from the base composition of the DNA. If the solution is then cooled rapidly, DNA remains single stranded but if it is allowed to cool slowly, reassociation occurs, and hyperchromicity is regained.

With preparations of DNA which were monodisperse in the ultracentrifuge, there was an abrupt rise in absorbtion over a small temperature range, the midpoint of which (T_m) was characteristic of the base composition of the DNA. With three H-bonds between GC base pairs, rather than the two with AT, the GC-rich DNA from *Micrococcus lysodeikticus* (72% G+C) had a much higher T_m than crab satellite poly (AT). Preparations of DNA from eukaryotes showed much broader curves consistent with their greater molecular complexity.

An important advance was made by Rich (1960), who showed it was possible to form double-stranded, H-bonded structures between complementary polyribonucleotide and polydeoxyribonucleotide chains. Schildkraut and co-workers demonstrated that duplex molecules could be formed between DNA of different microbial species. Filter techniques were developed (see Walker, 1969) which selectively retained these paired molecules. Usually the DNA was subjected to controlled shearing to reduce the length of the helices before heat denaturation to give single-stranded DNA. Labeled DNA or RNA molecules were annealed with the DNA under very carefully controlled conditions of temperature, ionic strength, and pH (often 0.15 M NaCl/0.015 M Na citrate, pH 7.0) (see McCarthy and Church,

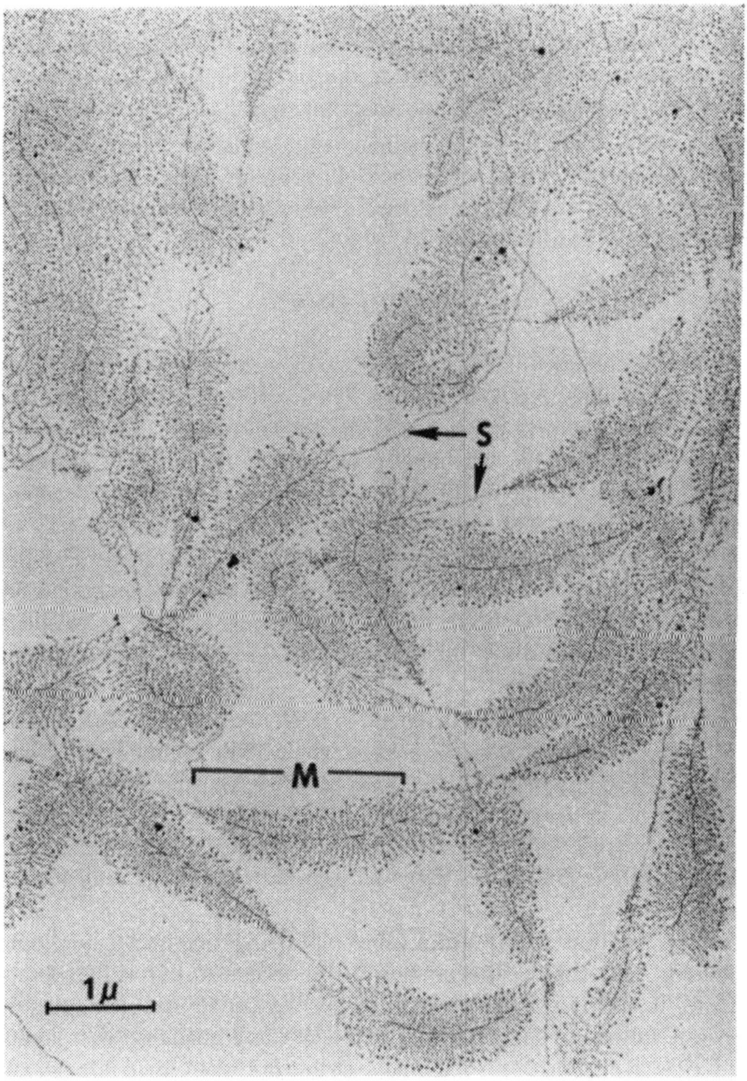

Figure 1. The transcription of nucleolar rRNA genes. S = untranscribed spacer DNA. M = matrix showing newly transcribed RNA molecules with bound protein. Tips of arrowheads indicate initiation points of RNA transcription. Reproduced, with permission, from Miller and Beatty, 1969.

1970). Complexes containing complementary DNA or RNA were retained on the filters.

Procedures were also developed for hybridization on cytological preparations *in situ*. In 1959, Kleinschmidt introduced a technique for preparing duplex or single-stranded DNA for electron microscopy using films spread on the air/water interface (see Kleinschmidt, 1968). With this method, regions of nonhomology were visualized in heteroduplexes. Nucleolar cores from Triturus oocytes showed fern-like figures with a central DNA fiber, and newly transcribed RNA molecules appearing as fronds (Figure 1) (Miller and Beatty, 1969).

Repetitive DNA

Hybridization studies with eukaryotic DNA showed some very singular results (see Britten and Kohne, 1969). With mouse DNA, c.10% reassociated very rapidly, and about 70% very slowly, as might be expected with single-copy genes. Mouse satellite DNA behaved like the rapidly associating fraction. C_0t curves were constructed relating the fraction of DNA reassociating to its initial concentration (C_0) and time (in secs). Sequences which occurred many times in the DNA, like satellite DNA, had low C_0t values and reassociated very much faster than unique or nearly unique regions. Comparisons of C_0t curves indicated the proportion of repetitive sequences in DNA from different species. Eukaryotic DNA had intermediate C_0t values, indicating there might be 100–10,000 repeated sequences containing several thousand residues.

Ribosomal RNA hybridized with repetitive DNA in eukaryotes; *Drosophila* had about 130 copies of rRNA genes, and *Xenopus* about 2000. Multiple copies of genes for tRNA were also indicated. Using *Xenopus* oocytes, Gall and Pardue (1969) and Birnstiel and his colleagues (John et al., 1969) showed rRNA hybridized to extrachromosomal rRNA genes (see below). Repetitive DNA was also located at centromeres (Pardue and Gall, 1969), and later, at telomeric ends of metaphase chromosomes.

In *Xenopus* additional rRNA genes were amplified during oocyte development. The additional copies were extrachromosomal and were lost during the subsequent progress of the embryo. Other repetitive genes were soon identified, notably for histones (see further in Stark and Wahl, 1984). It was argued that these additional copies of rRNA and tRNA were required to enable the organism to respond rapidly to conditions favorable for growth. These genes, in contrast to those used for protein synthesis, had only one stage of multiplication (see Orgel and Crick, 1980). The provision of new histones is essential for ongoing DNA replication. Multiple copies of histone genes ensure sufficiently rapid synthesis of the proteins in S phase (see Britten and Kohne, 1968). Other cases of gene amplification are now known—for example in instances of drug resistance. Here genes programming the synthesis of enzymes causing drug inactivation become amplified and are carried extrachromosomally or are permanently perpetuated within the genome (see Schimke, 1980).

It has been known since the 1940s that there is a rough correlation between amounts of DNA and the complexity of the organism. Major discrepancies, such as the excessive quantities of DNA in lilies and salamanders (ca. × 20 that in the human genome), provoked intense speculation which was enhanced by the discovery of repetitive DNA with low C_0t values (ca. 10^{-3}) compared to those for single-copy DNA (ca. 10^4).

This *C*-value paradox was particularly addressed by Britten (Britten and Kohne, 1968; Britten and Davidson, 1969) and by Walker and his associates (1969). Various suggestions were made—the repetitive sequences might be regulatory, controlling the expression of the genome. Alternatively or additionally repetitive DNA might be structural. Multiple copies of sequences might also have a protective role, allowing deletions or mutations to occur without damage to the organism; for example during aging, or permitting changes which lead to the acquisition of new functions.

In 1976, Richard Dawkin's book, *The Selfish Gene*, was published with the title immortalizing the phrase and superficially supporting the extreme view that organisms exist solely for the propagation of DNA, in spite of the need to ensure perpetuation of the organism the gene inhabits (Doolittle and Sapienza, 1980). Selfish DNA was considered to arise when a DNA sequence spread by forming additional copies of itself within the genome, but made no contribution to the genome (Orgel and Crick, 1980). More information is now available about the nature and sequences of satellite and repetitive DNA, and about smaller sequences repeatedly involved in gene regulation. However, the amount of DNA required for the latter function and for the proper structure of the centromere and telomeres is still uncertain. The jury is still out on repetitive DNA.

POLYMERASES AND RELATED ENZYMES

DNA Polymerases

The isolation of DNA polymerase from *E. coli* by A. Kornberg has already been mentioned. The protein, now called DNA pol I, has been intensively studied. In addition to its DNA synthesizing ability, it catalyzes two exonuclease activities, removing nucleotide bases $3' \rightarrow 5'$ and $5' \rightarrow 3'$. This latter activity is lost after limited proteolysis with trypsin or subtilisin, leaving the polymerase and 3′-5′ exonuclease activities only slightly diminished (Klenow and Henningsen, 1970). The two exonuclease activities were distinguished since the 5′-3′ nuclease preferentially utilizes double-stranded (ds) DNA, and can excise thymine dimers, which arise through the effects on DNA of UV irradiation. 3′-5′ exonuclease activity is arrested by the presence of dimers, and preferentially uses single-stranded DNA (see Goulian, 1971). The 3′-5′ nuclease is thought to be important for the removal of mismatched bases—proof-reading—so increasing the fidelity of DNA replication.

There was a serious problem respecting DNA pol I; the rate at which it synthesized DNA was only c.1% that observed *in vivo* (Kornberg, 1969). Many attempts were made to isolate more active preparations from the rapidly sedimenting, membrane-containing(?), fraction from *E. coli* extracts. These had faster rates of dXTP incorporation, but were only active for very brief periods (see Lark, 1969; Gefter, 1975).

In 1969 de Lucia and Cairns reported the isolation from *E. coli* of a nitrosoguanidine induced, temperature sensitive (t_s), DNA pol I mutant, which was apparently normal at 25–30°C, but was unable to replicate DNA at 45°C. At this temperature, however, it was able to perform repair synthesis. Cairns therefore concluded DNA pol I was not the enzyme primarily involved in DNA replication. Other t_s mutants were then examined. Those carrying mutations in the dnaE gene were found to have normal polymerases I and II but were unable to replicate at 42°C. At 30°C, an additional polymerase, III, was active, and was separated from the others by chromatography on phosphocellulose (see Gefter, 1975). DNA pol III was extensively purified by Otto et al. (1973). It had 3′-5′ exonuclease activity, was able to complement DNA synthesis in dnaE mutants, and incorporated deoxynucleotides at ca. 5×10^4 nucleotides/min, as *in vivo*. There are believed to be about 400 molecules of DNA pol I/cell in *E. coli*, but only ca. 10 of DNA pol III.

In 1963 Cairns used autoradiography and electron microscopy to examine ^{3}H-TdR uptake into DNA. The *E. coli* chromosome appeared to be a closed circular duplex, with 70–90 nm DNA. Replication proceeded from a fork, the limbs of which contained one old and one new strand (Figure 2). The presence of a replicating fork immediately presented a problem. DNA polymerases add deoxynucleotides 5′ → 3′. No enzyme was found operating in the 3′ → 5′ direction, which would be required for the complementary strand.

Evidence for discontinuous synthesis of DNA on the lagging strand (3′ → 5′) was forthcoming from Sakaba and Okazaki (1966). Using *E. coli* which had been cultured on ^{14}C-TdR, 10 s pulses of ^{3}H-TdR were used to identify very newly synthesized material. DNA was sedimented on alkaline sucrose gradients to separate its strands. Some very short oligodeoxynucleotides were found. Okazaki therefore proposed that replication in the 3′ → 5′ direction was achieved by synthesizing short lengths of 5′ → 3′ and joining them.

The need for a mechanism to join DNA fragments became apparent also from recombinant studies and from experiments with phage λ which has circular DNA. An enzyme was detected in *E. coli* infected with phage λ which had the capacity to join linear DNA covalently to give the mature, circular form (see Gefter, 1975). The enzyme, a ligase from *E. coli*, required NAD^+ for the reaction; that in T4 and T7 phages used ATP, as did the ligases later isolated from eukaryotes.

One final requirement for DNA replication is the need for a primer. Attachment of the entering dXTP is to a 3′ OH. Various experiments indicated RNA might be involved in the initiation of DNA synthesis (see Lark, 1969), and provide the 3′ OH group. Schekman et al. (1974) found complex ribonucleotide and RNA polymerase

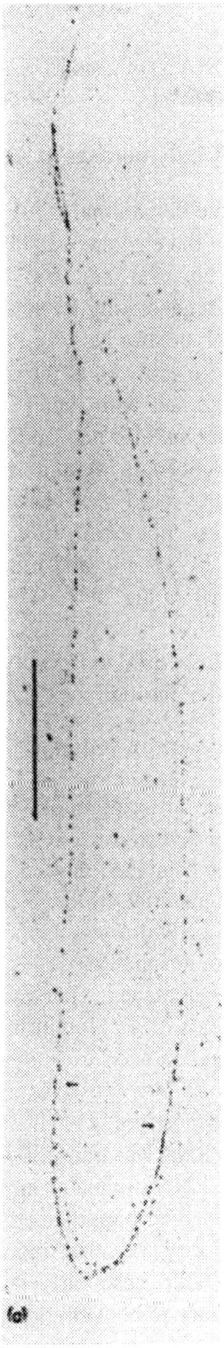

Figure 2. Autoradiograph of *E. coli* DNA following ^{3}HTdR incorporation. The arrows show the points of replication. Scale 100μ. Reproduced, with permission, from Cairns, 1963.

dependence for priming DNA synthesis in *E. coli*. Usually the priming ribonucleotide is subsequently removed.

DNA Polymerases in Animal Cells[a]

Bollum (1960) isolated the first animal DNA polymerase, DNA pol α, from calf thymus. Only about 25% of the enzyme was in the nucleus, and it had no $3' \rightarrow 5'$ exonuclease activity. It did however require a 3′ OH primer. Later studies, particularly with HeLa cells and regenerating liver, showed a second, smaller, β-polymerase to be present. β-polymerase activity is unchanged during the cell cycle, whereas the activity of the α-enzyme is increased in S phase. β-polymerase is thought to be mainly concerned with repair. As with *E. coli*, an RNA primer (Chargaff, 1976) and a ligase are also needed for replication.

A third polymerase, DNA pol γ, is found in mitochondria.

It is now known that many other enzymes and protein factors are required for replication in different viral, microbial, animal, and plant systems. They are gradually being identified by genetic and biochemical methods. A particularly important discovery was that of the DNA polymerase from the extreme thermophile, *Thermus aquaticus*, which is very resistant to heat denaturation (Saiki et al., 1988). The enzyme is therefore used in the polymerase chain reaction (PCR) (see Chapter 3) as it remains active through the denaturation/renaturation cycles.

RNA Polymerases

An enzyme catalyzing the transcription of RNA from a DNA template was first described by Weiss (1960). Polymerase activity was detected in crude preparations of liver nuclei. The enzyme catalyzed the incorporation of ^{32}P-labeled CTP and UTP into TCA-precipitated material, from which the radioactivity was released as mononucleotides after alkaline hydrolysis. Uptake into RNA required the presence of all four ribonucleoside triphosphates.

An intensive attack was then made to characterize the RNA polymerase activity in *E. coli*. Chamberlin and Berg (1962) found an enzyme with properties similar to those reported by Weiss, and showed it catalyzed net synthesis of RNA. When single-stranded DNA from øX174 was used as template, the base ratios in the RNA were in good agreement with those predicted from the DNA. When double-stranded (ds) øX174 DNA was used, RNA was transcribed from both strands. A final check established the newly made RNA stimulated amino acid uptake into protein in the *E. coli* ribosomal system. Several groups then obtained highly purified preparations of RNA polymerase from *E. coli* and other microorganisms (see Burdon, 1973).

A problem encountered with these early studies was that with ds DNA as template, RNA complementary to both strands was synthesized *in vitro*, whereas *in vivo* only one strand is copied. Experiments with circular DNAs such as øX174 showed that if the circles were intact, over 90% of the RNA was complementary to the mature strand of the phage, but if the circles were nicked, both strands were transcribed. Experiments with *Bacillus megatherium* infected with phage α gave similar results.

In eukaryotes the nuclear enzyme from ascites cells made single-stranded RNA. Its complementarity to the ascites DNA was confirmed by hybridization when the newly synthesized RNA was competed out by native ascites cell RNA; other RNAs were much less effective (see Burdon, 1973).

E. coli RNA polymerase did not require a primer. Maitra and Hurwitz (1965) used ^{32}P-(βγ)ATP and ^{32}P-(γ)UTP with poly d(AT) as template, to examine the direction of RNA synthesis and the fate of the initiating triphosphate. With labeled ATP, ^{32}P-phosphate was detected in the RNA, suggesting incoming XTPs condensed onto the 3′ OH of adenosine without loss of the terminal 5′ triphosphate. When uptake with ^{32}P-(γ)UTP was followed, very few chains were found to contain the labeled γ-P. As more DNAs were tested as templates, few RNA molecules were found to start with UTP or CTP, from which the authors concluded pyrimidine sites on double-stranded DNA were preferentially used to initiate RNA synthesis.

Further insight into the start of RNA transcription came in 1969 from a number of laboratories (see Burdon, 1973). When purified RNA polymerase from *E. coli* was chromatographed on phosphocellulose, it separated into a number of subunits. The minimal enzyme, which appeared to have the structure $\alpha_2\beta_2$ (now $\alpha_2\beta\beta_1$), transcribed phage T4 DNA poorly, but when a further component, the σ-subunit, was added, transcriptional activity was restored. σ-factor did not affect the rate at which the ribonucleotides were elongated, but did promote initiation. When RNA synthesis by the holoenzyme was checked against the protein whose synthesis the RNA directed, σ-factor was shown to be required for transcription to be initiated from the appropriate start on the gene. The factor was subsequently released from the complex and could be reutilized for further initiation. Initiation was inhibited by rifamycins.

Only one RNA polymerase was detected in *E. coli*, whereas multiple RNA polymerases have been found in mammalian systems. Widnell and Tata (1964, 1966) prepared Weiss' aggregate enzyme from rat liver nuclei. Two different activities were detected. One—now RNA pol I—which was Mg^{2+}-dependent, was very sensitive to actinomycin D and made rRNA. It was also specifically inhibited by α-amanitin. A second enzyme was activated by Mn^{2+} and was less sensitive to actinomycin. It, RNA pol II, catalyzed the incorporation of ribonucleotides with a base ratio similar to that in total nuclear DNA rather than rDNA. Some of the features of mRNA transcription by RNA pol II have already been mentioned.

Later a third enzyme, RNA pol III, was found in eukaryotes, catalyzing the synthesis of 5s and t RNAs. Viral RNA polymerases and RNA-dependent RNA polymerases (replicases) are also known.

The viral polymerase which is essential for the multiplication of retroviruses, reverse transcriptase, uses its own strand of RNA as a template to make DNA. The existence of such an enzyme had been postulated by Temin[14] (1964) to explain why inhibitors of DNA synthesis, such as methotrexate, 5-fluorodeoxyuridine, and cytosine arabinoside, blocked the replication of the Rous sarcoma RNA virus. Temin proposed the replication of RNA tumor viruses took place through DNA intermediates—the DNA provirus hypothesis. Reverse transcriptases were isolated by Temin and Mizutani (1970) from Rous sarcoma virus (RSV) and by Baltimore (1970) from RSV and Rauscher mouse leukemic virus (MLV). Spiegelman et al. (1970) used separation on a Cs_2SO_4 gradient to demonstrate the existence of an intermediate RNA–DNA hybrid in the replication of Rauscher MLV. Further analysis revealed the complexities of these polymerases and the need for various associated factors for their activity *in vivo*.

Recognition of the existence of reverse transcriptase was tremendously important, first in understanding the propagation and spread of RNA tumor viruses, and how a viral infection could lead to oncogenic transformation through the integration of the virally programmed DNA into the host genome. Second, the ability of the enzyme to use RNAs of nonviral origin allowed mRNAs, which were known to program the synthesis of particular proteins, to be copied to yield cDNA. The finding that cDNAs hybridized to discontinuous regions of the genome led directly to the discovery of split genes in eukaryotes, introns, and exons, and thus to gene splicing (for refs., see Gilbert, 1978; Crick, 1979). (Introns are transcribed DNA sequences which intervene between exons and have to be excised. The exons are joined up to form the structural gene.)

The existence of reverse transcriptase appeared to conflict with textbook formulations of the Central Dogma: Information flowed from DNA to RNA to protein. As we have already recounted, the original formulation by Crick (1958) only excluded protein → protein and protein → nucleic acid information transfer. "I have never suggested it [the transfer of information from RNA to DNA] cannot occur." (Crick, 1970).

Restriction Enzymes

One further class of enzymes affecting DNA must be mentioned—restriction enzymes—endonucleases which hydrolyze DNA at specific deoxynucleotide sequences. In 1962, Arber and Dussoix observed marked differences in the capacity of phage λ to proliferate in different strains of *E. coli*—*E. coli* K12 and *E. coli* B. Propagation of the phage depended on the presence of S-adenosylmethionine, a prerequisite for enzymic methylation. They found that if the phage could be methylated within its host, its DNA was protected from endonuclease attack, and phage multiplication followed. Unmethylated DNA was degraded and phage propagation prevented (see Arber and Linn, 1969).

This observation stimulated numerous studies by which different classes of restriction enzymes were recognized, and the sequence specificities, defined by four or more bases, identified. Class II restriction enzymes only have endonuclease activity. DNA methylation is performed by a separate enzyme. Class II enzymes are essential to yield the overlapping base sequences necessary for DNA sequencing (Sanger et al., 1977; Maxam and Gilbert, 1980).

SUMMARY

From 20 to 25 years elapsed between the publication of the double helical structure for DNA and the start of the molecular biology revolution. Biochemistry was not dormant during that period. The methods of information storage in DNA and its transfer via RNA to protein synthesis were established, mainly by standard biochemical procedures. These were supplemented by genetic recombination analysis and the effective use of DNA and RNA viruses. All the data were acquired before DNA sequences were determined. Details of the mechanisms of replication, transcription, and translation had still to be uncovered, especially requirements for numerous accessory proteins interacting with DNA or RNA, many of whose roles were nonenzymic. Classical biochemical studies during the same period led to the discovery of reverse transcriptase and the restriction enzymes, which, together with the ability to sequence genes and the development of more efficient procedures for cell transformation, were to be the principle tools for the detailed analysis and exploitation of molecular genetics now in progress.

NOTES

[a]See also Weissbach, 1975.

REFERENCES

Abrams, R. (1961). Nucleic acid metabolism and biosynthesis. Annu. Rev. Biochem. 30, 165–188.

Arber, W. & Linn, S. (1969). DNA modification and restriction. Annu. Rev. Biochem. 38, 467–500.

Avery, O.T., McLeod, C.M., & McCarty, M. (1944). Studies on the chemical nature of the substance inducing transformation of Pneumococcal types. J. Exp. Med. 79, 137–158.

Baltimore, D. (1970). Viral RNA-dependent DNA polymerase. Nature 226, 1209–1211.

Beadle, G.W. (1945). Genetics and metabolism in Neurospora. Physiol. Rev. 25, 643–663.

Bollum, F.J. (1960). Calf thymus polymerases. J. Biol. Chem. 235, 2399–2403.

Brachet, J. (1957). Biochemical Cytology. Academic Press, New York.

Brawerman, E. (1974). Eukaryotic mRNA. Annu. Rev. Biochem. 43, 621–642.

Brenner, S., Jacob, F., & Meselson, H. (1961). An unstable intermediate carrying information from genes to ribosomes for protein synthesis. Nature 190, 576–581.

Britten, R.J. & Davidson, E.H. (1969). Gene regulation for higher cells: A theory. Science 165, 349–357.

Britten, R.J. & Kohne, D.E. (1968). Repeated sequences in DNA. Science 161, 529–540.

Britten, R.J. & Kohne, D.E. (1969). Repetition of nucleotide sequences in chromosomal DNA. In: Handbook of Molecular Cytology (Lima di Faria, A., Ed.), pp. 21–51. North Holland Publishing, Amsterdam.

Brown, D.D. & Gurdon, J.B. (1964). Absence of ribosomal RNA synthesis in the anucleolate mutant of *Xenopus laevis*. Proc. Natl. Acad. Sci. USA 51, 139–146.

Burdon, R.H. (1973). Nucleic acid biosynthesis and interactions. In: Cell Biology in Medicine. (Bittar, E.E., Ed.), pp. 280–324. Wiley & Sons, New York.

Cairns, J. (1963). The bacterial chromosome and its manner of replication as seen by autoradiography. J. Mol. Biol. 6, 208–213.

Caspersson, T. (1950). Cell Growth and Function. Norton, New York.

Chamberlin, M. & Berg, P. (1962). DNA-directed synthesis of RNA by an enzyme from *E. coli*. Proc. Natl. Acad. Sci. USA 48, 81–94.

Chantrenne, H., Burny, A., & Marbaix, G. (1967). The search for the mRNA for hemoglobin. Prog. Nucleic Acids Res. & Molec. Biol. 7, 173–194.

Chargaff, E. (1976). Initiation by RNA of DNA synthesis. Prog. Nucleic Acids Res. & Molec. Biol. 16, 1–24.

Chargaff, E. & Davidson, J.N. (1955–1960). The Nucleic Acids., Vols. I–III. Academic Press, New York.

Crick, F.H.C. (1958). On protein synthesis. Soc. Exp. Biol. Symposium 12, 138–163.

Crick, F.H.C. (1963). Recent excitement in the coding problem. Prog. Nucleic Acids Res. & Molec. Biol. 1, 163–217.

Crick, F.H.C. (1966). Codon:anticodon pairing. The wobble hypothesis. J. Mol. Biol. 19, 548–555.

Crick, F.H.C. (1970). The central dogma of molecular biology. Nature 227, 561–563.

Crick, F.H.C. (1979). Split genes and RNA splicing. Science 204, 264–271.

Crick, F.H.C., Griffith, J.S., & Orgel, L.E. (1957). Codes without commas. Proc. Natl. Acad. Sci. USA 43, 416–421.

Darnell, J.E. (1976). mRNA structure and function. Prog. Nucleic Acids Res. & Molec. Biol. 19, 376–511.

Dawkins, R. (1976). The Selfish Gene. Oxford University Press.

Dintzis, H.M. (1961). Assembly of the peptide chains of hemoglobin. Proc. Natl. Acad. Sci. USA 47, 247–261.

Doolittle, W.F. & Sapienza, C. (1980). Selfish genes, the phenotype paradigm and genomic evolution. Nature 284, 601–603.

Edmonds, M. & Caramela, M.G. (1969). The isolation and characterization of adenosine monophosphate-rich polynucleotides synthesized by Erlich ascites cells. J. Biol. Chem. 244, 1314–1324.

Fraenkel-Conrat, H. & Singer, B.A. (1957). Virus reconstitution: Combination of protein and nucleic acid from different strains. Biochim. Biophys. Acta 24, 540–548.

Franklin, R.E. & Gosling, R.G. (1953). Molecular configuration of sodium thymonucleate. Nature, Lond. 171, 740–741.

Gall, J.G. & Pardue, M.L. (1969). Formation and detection of RNA–DNA hybrid molecules in cytological preparations. Proc. Natl. Acad. Sci. USA 63, 378–383.

Gamow, G. (1954). Possible relation between DNA and protein structure. Nature 173, 318.

Gefter, M.L. (1975). DNA replication. Annu. Rev. Biochem. 44, 45–78.

Gierer, A. & Schramm, G. (1956). Infectivity of RNA from tobacco mosaic virus. Nature 177, 702–703.

Gilbert, W. (1978). Why genes in pieces? Nature 271, 501.

Goulian, M. (1971). Biosynthesis of DNA. Annu. Rev. Biochem. 40, 855–898.

Griffith, F. (1928). Significance of Pneumococcal types. J. Hyg. (Camb.) 27, 113–159.

Grunberg-Manago, M. (1963). Polynucleotide phosphorylase. Prog. Nucleic Acids Res. & Molec. Biol. 1, 93–133.

Hammerling, J. (1953). Nucleo-cytoplasmic relationships in the development of acetobularia. Intern. Rev. Cytol. 2, 475–498.

Harris, H. (1968). Nucleus and Cytoplasm. Clarendon Press, Oxford, UK.

Hershey, A.D. & Chase, M. (1952). Independent functions of viral protein and nucleic acid in growth of bacteriophage. J. Gen. Physiol. 36, 39–59.

Holley, R.W. (1968). Experimental approaches to the determination of the nucleotide sequences of large oligonucleotides and small nucleic acids. Prog. Nucleic Acids Res. & Molec. Biol. 8, 37–47.

Jacob, F. & Monod, J. (1961). Genetic regulatory mechanisms in the synthesis of proteins. J. Mol. Biol. 3, 318–356.

John, H.A., Birnstiel, M.L., & Jones, K.W. (1969). RNA–DNA hybrids at the cytological level. Nature 223, 582–587.

Judson, H.F., Ed. (1979). Eighth day of Creation. Simon & Schuster, New York.

Khorana, H.G. (1959). Synthesis and structural analysis of polynucleotides. J. Cell. Comp. Physiol. 54, suppl. 1, 5–15.

Kit, S. (1963). Deoxyribonucleic acid. Annu. Rev. Biochem. 39, 131–150.

Kleinschmidt, A.K. (1968). Monolayer techniques in electron microscopy of nucleic acids. Methods in Enzymology B 12, 361–377.

Klenow, H. & Henningsen, I. (1970). Selective elimination of the exonuclease activity of DNA polymerase from *E. coli* by limited proteolysis. Proc. Natl. Acad. Sci. USA 65, 168–175.

Kornberg, A. (1968). DNA Replication. W.H. Freeman, New York.

Kornberg, A. (1969). The active center of DNA polymerase. Science 163, 1410–1418.

Lane, C.D., Marbaix, G., & Gurdon, J.B. (1971). Rabbit hemoglobin synthesis in frog cells; the translation of reticulocyte 9s RNA in frog oocytes. J. Mol. Biol. 61, 73–91.

Lark, K.G. (1969). Initiation and control of DNA synthesis. Annu. Rev. Biochem. 38, 569–604.

Lengyel, P., Streyer, J.F., & Ochoa, S. (1961). Synthetic polynucleotides and the amino acid code. Proc. Natl. Acad. Sci. USA 47, 1936–1942.

Lipmann, F. (1941). The metabolism, generation and utilization of phosphate bond energy. Adv. Enzymol. 1, 99–162.

Lucia, P. de & Cairns, J. (1969). Isolation of an *E. coli* strain with a mutation affecting DNA polymerase. Nature 224, 1164–1168.

Maitra, U. & Hurwitz, J. (1965). The role of DNA in RNA synthesis: Nucleoside triphosphate terminii in RNA polymerase products. Proc. Natl. Acad. Sci. USA 54, 815–822.

Marker, K., Clark, B.F.C., & Anderson, J. (1966). *N*-formyl-methionyl sRNA and its relation to protein synthesis. Cold Spring Harbor Symposium on Quantitative Biology 31, 279–285.

Matthei, J.H. & Nirenberg, M.W. (1961). Dependence of cell-free protein synthesis in *E. coli* upon naturally occurring or synthetic polyribonucleotides. Proc. Natl. Acad. Sci. USA 47, 1580–1588; 1588–1602.

Maxam, A.M. & Gilbert, W. (1980). A new method for sequencing DNA. Proc. Natl. Acad. Sci. USA 74, 560–564.

McCarthy, B.J. & Church, R.B. (1970). The specificity of molecular hybridization reactions. Annu. Rev. Biochem. 39, 131–150.

Meselson, M. & Stahl, F.W. (1958). Semi-conservative replication in *E. coli*. Proc. Natl. Acad. Sci. USA 44, 671–682.

Miller, O.L. & Beatty, B.R. (1969). Portrait of a gene. J. Cell. Physiol. 74, suppl. 1, 225–232.

Nirenberg, M.W. & Leder, P. (1964). RNA codewords and protein synthesis. Proc. Natl. Acad. Sci. USA 52, 420–427.

Ord, M.G. & Stocken, L.A. (1995). Early Adventures in Biochemistry. Foundations of Modern Biochemistry. Vol. 1. JAI Press, Greenwich, CT.

Orgel, L.E. & Crick, F.H.C. (1980). Selfish DNA: The ultimate parasite. Nature 284, 604–607.

Otto, B., Bonhoeffer, F., & Schaller, H. (1973). Purification and properties of DNA polymerase III. Eur. J. Biochem. 34, 440–447.

Pardee, A.B. (1985). Molecular basis of gene expression: Origins from the pajama experiment. BioEssays 2, 86–89.

Pardue, M.L. & Gall, J.G. (1969). Molecular hybridization of radioactive DNA to the DNA of cytological preparations. Proc. Natl. Acad. Sci. USA 64, 600–604.

Pauling, L. & Corey, R.B. (1953). A proposed structure for the nucleic acids. Proc. Natl. Acad. Sci. USA 39, 84–97.

Perry, R.P. (1969). Nucleoli—the cellular site for ribosome production. In: Handbook for Molecular Cytology (Lima di Faria, A., Ed.), pp. 620–636. North Holland Publishing, Amsterdam.

Rich, A. (1960). A hybrid helix containing both deoxyribose and ribose polynucleotides and its relation to the transfer of information between the nucleic acids. Proc. Natl. Acad. Sci. USA 46, 1044–1053.

Saiki, R.K., Gelfand, D.H., Stoffel, S., Scharf, S.J., Higuchi, R., Horn, G.T., Mullis, K.B., & Erlich, H.A. (1988). Primer-directed enzymatic amplification of DNA with a thermostable DNA polymerase. Science 239, 487–491.

Sakaba, K. & Okazaki, R. (1966). A unique property of the replicating region of chromosomal DNA. Biochim. Biophys. Acta 129, 651–654.

Sanger, F. (1952). Arrangements of amino acids in proteins. Adv. Prot. Chem. 7, 1–67.

Sanger, F., Nicklen, S., & Coulson, A.R. (1977). DNA sequencing with chain-terminating inhibitors. Proc. Natl. Acad. Sci. USA 74, 5463–5467.

Sayre, A. (1975). Rosalind Franklin and DNA. W.W. Norton, New York, London.

Schekman, R., Weiler, A., & Kornberg, A. (1974). The need for an RNA primer for DNA synthesis. Science 186, 987–993.

Schimke, R.T. (1980). Gene amplification and drug resistance. Sci. Amer. 243, 60–69.

Siekevitz, P. & Palade, G.E. (1960). A cytochemical study of the pancreas of the guinea pig. J. Biophys. Biochem. Cytol. 7, 619–630; 631–644.

Spiegelman, S., Burny, A., Das, M.R., Keyder, J., Schlam, J., Travnicek, M., & Watson, K. (1970). Characterization of the products of RNA-induced DNA polymerase in oncogenic RNA viruses. Nature 227, 563–567.

Stark, G.R. & Wahl, G.M. (1984). Gene amplification. Annu. Rev. Biochem. 53, 447–491.

Stent, G.S. (1971). Molecular Genetics. W.H. Freeman, New York.

Taylor, J.H., Woods, P.S., & Hughes, W.L. (1957). The replication of DNA in *E. coli*. Proc. Natl. Acad. Sci. USA 43, 122–128.

Temin, H.M. (1964). Carcinogenesis by avian sarcoma virus. Cancer Res. 28, 1835–1838.

Temin, H.M. & Mizutani, S. (1970). RNA-dependent DNA polymerase in virions of Rous sarcoma virus. Nature 226, 1211–1213.

Volkin, E. & Astrachan, L. (1956). ^{32}P incorporation in *E. coli* RNA after infection with bacteriophage T2. Virology 2, 146–161.

Walker, P.M.B. (1969). The specificity of molecular hybridization in relation to studies on higher organisms. Prog. Nucleic Acids Res. & Molec. Biol. 9, 301–326.

Watson, J.D. (1965). Molecular Biology of the Gene. 1st ed. W.A. Benjamin, New York, Amsterdam.

Watson, J.D. (1968). The Double Helix. Weidenfeld & Nicolson, London.

Watson, J.D. & Crick, F.H.C. (1953). A structure for deoxyribose nucleic acid. Nature 171, 737–738.

Weiss, S.B. (1960). Enzymic incorporation of ribonucleoside triphosphates into interpolynucleotide linkage of RNA. Proc. Natl. Acad. Sci. USA 46, 1020–1030.

Weissbach, A. (1975). Vertebrate DNA polymerases. Cell 5, 101–108.

Widnell, C.C. & Tata, J.R. (1964). Evidence for two DNA-dependent RNA polymerase activities in isolated rat liver nuclei. Biochim. Biophys. Acta 87, 531–533; (1966). 123, 478–492.

Wilkins, M.H.F., Stokes, A.R., & Wilson, H.R. (1953). Molecular structure of deoxypentose nucleic acids. Nature, Lond. 171, 738–740.

Woese, C.R. (1967). Present status of the genetic code. Prog. Nucleic Acids Res. & Molec. Biol. 7, 107–172.

Yanofsky, C. (1967). Gene structure and protein structure. Sci. Amer. 216, 80–94.

Chapter 3

MANIPULATING DNA:
FROM CLONING TO KNOCKOUTS

Jan A. Witkowski

INTRODUCTION

There are times when a field of scientific research enters a period of consolidation when the outstanding old problems have been worked out; research turns to filling in the details—dotting the i's and crossing the t's in the established framework rather than tackling new and novel problems. This is in part because these new problems are insoluble with the current concepts and techniques. Some radical change is needed to open up a fresh set of problems to experimental investigation.

Molecular genetics was in such a period in the late 1960s. (The book by Judson, 1978, gives an interesting account of post-double helix research with many comments from the participants.) The flowering of bacterial and phage genetics that followed elucidation of the double helical structure of DNA had revealed the intimate workings of genes in these organisms. Andre Lwoff, Joshua Lederberg, and Norton Zinder had exploited gene exchange between bacteria; Seymour Benzer had "run the map into the ground" with his fine mapping of the rII region; Jacques Monod and Francois Jacob laid the foundations of gene regulation with their studies of the *lac* operon; and proteins that interact with and control gene expression were found by Walter Gilbert and Benno Muller-Hill (*lac* repressor) and Mark Ptashne (λ phage repressor).

The mechanism of protein synthesis came under intense experimental attack in the late 1950s and through the 1960s. Crick's adaptor hypothesis postulated the existence of the transfer RNAs that were found (independently of the prediction) by Zamecnik,[28] Hoagland, and their colleagues (Hoagland, 1990). Messenger RNA had been sighted by Eliot Volkin and L. Astrachan as early as 1956 (Volkin and Astrachan, 1956; Volkin, 1995) and their findings were confirmed using different techniques (Hall and Spiegelman, 1960) and different organisms (Ycas and Vincent, 1960). Then, data from classical experiments performed at the Californian Institute of Technology (Brenner et al., 1961) and at Harvard (Gros et al., 1961) showed there was a short-lived RNA with properties expected for a messenger RNA. The ribosome was taken apart and put together again; the mechanism by which activated amino acids become attached to their transfer RNAs was determined; and the biochemical details of protein synthesis worked out (see Chapter 5, and Watson et al., 1987).

The holy grail of that time was deciphering the genetic code but the initial theoretical analyses (that on occasion ignored biology) were rather unsatisfactory. Then Brenner showed that the code had to be non-overlapping and Crick demonstrated that the code was almost certainly one in which triplets of nucleotides coded for amino acids. But it was through the use of straightforward and clever biochemical means that the codon assignments were made, much to the surprise of many who had expected the problem to be intractable.

The progress was such that by 1966, Crick[5] felt that ". . . the *foundations* (his italics) of molecular biology were now sufficiently firmly outlined that they could be used as a fairly secure basis for the prolonged task of filling in the many details" (Crick, 1988). Even the genetic code—the Rosetta Stone to understanding the "secrets of life"—led nowhere because, if the code words could now be deciphered, they could not be read. The problems of determining the sequence of nucleotides in an RNA molecule was difficult—in a DNA molecule, next to impossible. It was at about this time that Gunter Stent proclaimed the end of molecular biology (1968) and he, Crick, Brenner, and Benzer looked for new challenges in other fields. Brenner sought out a new experimental organism and with help from his friends established the nematode *C. elegans* as the new *Drosophila* (Brenner, 1988; Wood, 1988). But the playing field of molecular biology was about to change. As Crick wrote in his autobiography, "Although I did not appreciate it, molecular biology was on the verge of a massive step forward, caused by three new techniques: recombinant DNA, rapid DNA sequencing, and monoclonal antibodies" (Crick, 1988).

RECOMBINANT DNA

I am going to concentrate on the contributions of DNA to that "massive step," contributions that fall under the general rubric of "recombinant DNA" that has come to encompass experimental manipulations of both DNA and RNA together with related techniques such as expressing proteins in bacteria. Recombinant DNA methods are used in almost all areas of experimental biology, from analyzing the atomic contacts between proteins and nucleic acids, to DNA fingerprinting in animal ecology. It came to play such an important role because the techniques made possible new approaches in three major areas: making DNA; analyzing DNA; and studying the functions of DNA. That is, recombinant DNA provided the means to study genes.

There are so many topics that I can select only a small set and, in an essay of such broad scope, none can be examined in detail. (A much more detailed description of many more topics will be found in Watson[7] et al., 1992. The book by Winnacker, 1987, provides technical descriptions of the earlier techniques.) I regret especially that there is not room to review the extraordinary contribution computers are making to molecular biology nor to discuss the genome projects, arguably the current research most likely to have far-reaching consequences. Furthermore, the picture that will emerge is one continuous success as I move from landmark to landmark. This is not a true picture of research, as historians and sociologists of science never fail to emphasize to scientists who write the history of their own research field. In fact, scientists understand this very well, being the ones who have the most immediate experience of both the failures and successes of research.

I am going to begin in 1970 when several key steps were taken, one of which, the discovery of reverse transcriptase, in retrospect can be seen to herald the coming revolution.

MAKING GENES AND DNA

During the 1960s detailed genetic analysis could be performed on viruses and bacteria, but the genomes of higher organisms were too large and complex to analyze at the molecular level. Being able to isolate genes and to make copies of them by cloning was a prerequisite for the development of molecular genetics. With a gene isolated and available in large amounts, its structure and function can be analyzed directly in contrast to "classical" genetics that relies on inferences from observable phenotypes. Cloning is not the only way of making DNA and chemical methods had been developed long before. However, it took further developments in the early 1980s for chemical DNA synthesis to become a routine tool for molecular applications. The third method for making DNA that I shall describe is the polymerase chain reaction. This technique is extraordinarily versatile and has brought about a revolution in DNA and genetic manipulations.

Cloning

The Enzymes

Throughout the 1950s and 1960s, biochemists had been isolating enzymes that acted upon DNA. Most notable were the DNA polymerases, the first of which, DNA polymerase I, was isolated by Arthur Kornberg[9] in 1956. The DNA polymerases and the RNA polymerases were a sensation for biochemists. Kornberg quotes a conversation that he had with Joseph Fruton in which Fruton[25] expressed great scepticism of Kornberg's results because: "There is no known case in which an enzyme takes instructions from its substrate" (Kornberg, 1989).

This was not the only doctrine to fall by the wayside. It was one of the foundations of molecular genetics that information moved from DNA→RNA→Protein. But what was to be made of the retroviruses that had a viral RNA genome but integrated into the host cell DNA in a proviral DNA form? Howard Temin[14] suggested that there had to be an enzyme capable of synthesizing DNA on an RNA template, but this was unacceptable according to the prevailing "Central Dogma". However, in 1970, Temin and Mizutani (1970), and David Baltimore (1970), independently discovered the enzyme reverse transcriptase that does precisely that. The enzyme works also using eukaryotic mRNA as a substrate and so it became possible to produce a DNA copy of a given mRNA, a so-called copy DNA or cDNA.

By 1970, other enzymes including the DNA ligases that join DNA strands end-to-end, terminal transferase that adds nucleotides to the 3′ ends of DNA molecules, exonucleases that remove nucleotides, methylases that add methyl groups to nucleotides in a DNA chain, and restriction endonucleases that cut DNA, had been isolated. The restriction enzymes were an unexpected discovery (Arber and Linn, 1969; Arber, 1979) and a curiosity until Smith and Wilcox (1970) (Smith, 1979) showed that one class of restriction endonucleases, type II, have substrate specificity that is defined by the sequence of nucleotides that each enzyme recognizes. For example, the enzyme *Eco*RI cuts double-stranded DNA only at the sequence GAATTC. Now those studying nucleic acids had enzymes of a specificity that surpassed the proteases that had been used so successfully by protein chemists. All these enzymes constituted the tools that were soon to be used in genetic engineering.

Recombining DNA

The first recombining of DNA fragments from different sources was reported by Jackson et al. (1972) (Berg, 1981). A key element of their approach was the use of a technique that appeared in the *Journal of Molecular Biology* one year after the Jackson et al. paper was published. Lobban and Kaiser (1973) showed that it was possible to convert linear phage 22 DNA into double-stranded circles, closed covalently. This they did by first using λ phage exonuclease to produce 3′ single-strand ends on the phage DNA; terminal transferase was used to add strings of adenines to one preparation and strings of thymidines to another; annealing the two populations of molecules through these poly-A and poly-T extensions; and finally treating with DNA ligase, exonuclease III, and DNA polymerase I to covalently link and repair the annealed strands.

The long-term aim of the Berg laboratory was to devise a means ". . . by which new, functionally defined segments of genetic information" can be introduced into mammalian cell. This dictated Berg's choice of a vector—a DNA tumor virus that infected mammalian cells with high efficiency. Jackson et al. made chimeric molecules of SV40 DNA and DNA from an *E. coli* plasmid that contained phage DNA and the complete *E. coli* galactose operon. While they were able to demonstrate, using physical means, that chimeric molecules had been formed, the biological activity of these chimeric molecules could not be tested—the *Eco*RI site used to linearize the circular SV40 genome cut a gene essential for replication. Further work on SV40 stopped because of concerns about the safety of using DNA from an oncogenic virus as a vector. These worries led to the Berg letter and to the great recombinant DNA debates of the 1970s (Rogers, 1977; Watson and Tooze, 1981).

DNA Cloning

The work that demonstrated the utility of recombinant DNA was that of Cohen,[17] Boyer, and their colleagues (Cohen et al., 1973; Cohen, 1988). There were two special features of their approach. The first was that Cohen had been studying antibiotic resistant plasmids and his group devised a method for introducing plasmids into *E. coli* and using the acquisition of antibiotic resistance to select for bacterial cells that had taken up the plasmids. A bacterial cell would give rise to a clone that contained descendants of the plasmid that had transformed the cell. Suppose, then, that different fragments of foreign DNA could be inserted in plasmids. Transformation would provide a means of producing unique clones of bacterial cells, each clone containing just one of those plasmids and one of those fragments of foreign DNA. The second was that Mertz and Davis (1972) had shown that *Eco*RI (isolated by Boyer) cut DNA so as to produce 3′ and 5′ complementary single strands that could be joined very efficiently by DNA ligase. The pSC101 plasmid has a single site for *Eco*RI and after cutting with *Eco*RI, the fragments were mixed with *Eco*RI fragments that contained a kanamycin-resistant gene derived from another plasmid, R6-5. After transformation into *E. coli*, bacterial clones were found that expressed resistance to tetracycline (on pSC101) *and* kanamycin (from R6-5).

The basic strategy for cloning any fragment of DNA was clear from these experiments. First, isolate DNA containing the gene of interest. Second, choose a suitable vector-DNA capable of replicating in a host cell. Third, prepare fragments of the DNA of a size to fit into the vector. Fourth, introduce these recombinant DNA molecules into bacterial cells and isolate clones. Fifth, identify the clones containing the gene or DNA of interest. The key features of cloning are that a specific DNA sequence can be isolated and that very large amounts of the DNA can be made.

Cloning Eukaryotic Genes-I: cDNA Libraries

While the immediate beneficiaries of cloning were scientists working on prokaryotes, scientists working on eukaryotic genes seized the opportunity to analyze these genes at a level that had been achieved previously only for viruses and bacteria. However, there were some serious difficulties to be overcome, stemming from the enormous complexity of eukaryotic genomes compared with those of viruses and bacteria. How was it possible to prepare cloned DNA fragments from eukaryotic genomes and to identify which clones contained the DNA of interest? The strategy that was followed was to reduce the complexity by using cells that made a great deal of a single protein and to use reverse transcriptase to make DNA copies of messenger RNAs in those cells. Obvious target genes for cloning were the globin genes. These were among the best characterized eukaryotic genes; α- and β-globin mRNAs could be purified readily from reticulocytes, and these mRNAs are major fractions of the mRNAs in these cells. Cloning these became the goal of several laboratories (Rougeon et al., 1975; Efstratiadis et al., 1976; Rabbitts, 1976) and I shall take the experiments of Efstratiadis, Maniatis,[19] and their colleagues as an example.

The full-length, single-stranded cDNA molecules made by reverse transcriptase have short double-stranded hairpins at their 3′ end. Efstratiadis et al. (1976) showed that these hairpins could act as primers for the synthesis of a complete second strand by DNA polymerase, producing full length double-stranded cDNA copies of rabbit globin. The logical next step was to take these molecules and clone them (Maniatis et al., 1976). First, treatment with S1 nuclease cut the single strand of the hairpins and degraded any single-stranded cDNA. Full length molecules were isolated by polyacrylamide gel electrophoresis. These molecules and the PMB9 plasmid vector derived from Cohen's pSC101 were treated with λ-exonuclease to cut back the 5′ ends. Strings of poly(dT) were added to the globin molecules while poly(dA) ends were added to the vector. The globin and plasmid molecules were ligated together, transformed into *E. coli*, and transformants selected by growing on tetracycline. The technique developed by Grunstein and Hogness (1975) enabled Maniatis et al. (1976) to screen colonies positive for β-globin. Much of the group's work was taken up with detailed restriction mapping of the clones to show that no rearrangements had occurred. This was the basic strategy until the early 1980s when the "replacement" method was devised by Okayama and Berg (1982). In this procedure, the mRNA–cDNA hybrid is treated with RNAase that produces gaps in the mRNA strand. The short stretches of mRNA then act as primers for DNA polymerase that synthesizes the second DNA strand.

There was an advantage to cDNA cloning that became apparent only with hindsight. In the same year cDNA cloning of the globin genes was reported, RNA splicing was discovered by Berget and Sharp at Massachusetts Institute of Technology (Berget et al., 1977) and by a group at Cold Spring Harbor Laboratory including Roberts,[18] Gelinas, Chow, and Broker (Chow et al., 1977). Comparisons between the adenovirus genome and mRNAs produced from it showed that the DNA contained sequences that were not present in the mRNAs. The DNA sequence of a gene was interrupted by sequences that did not code for amino acids; the coding portions of a gene were called exons and the noncoding sequences, introns. This, one of the most remarkable discoveries of the recombinant DNA era (Witkowski, 1988), had two important consequences for gene cloners. First, it became evident that bacteria could not remove introns from eukaryotic genes. If eukaryotic proteins were to be made in bacteria, the genetic information would have to be supplied as cDNAs. Second, interesting facts were likely to come from studying the organization and structure of eukaryotic genes but this was going to require cloning of genomic DNA. In addition, sequences involved in regulating gene expression lie outside the coding sequence and genomic DNA was essential for examining them.

Cloning Eukaryotic Genes-II: Genomic Libraries

Quite apart from these reasons for wanting to clone genomic DNA, there were problems with cDNA cloning. The most serious arose when trying to prepare cDNA clones of all the species of mRNAs in a tissue. Because cDNAs are made from mRNAs, the proportion of cDNA for a particular mRNA depends on the concentration of that mRNA in the cells used. This can be turned to one's advantage if trying to clone the gene for a protein that is produced in large amounts in a certain cell type. Those cells will be enriched for the mRNA for that protein and the cDNA library will be enriched for the corresponding cDNA. This is why the cDNAs for proteins such as globin, ovalbumin, and the chorion proteins of insects were the first targets for cDNA cloning. But it is very difficult to produce a library of cDNAs that has equal representation of all mRNAs, from the very abundant to the very rare—cDNA cloning is not the way to go if the aim is to clone *all* the genes in a cell.

Three major problems had to be overcome before genomic DNA cloning became possible. The first was political. As a consequence of the alarm raised about possible dangers of recombinant DNA experiments (Watson and Tooze, 1981), and following the famous Asilomar Conference in 1975 (Rogers, 1977), certain conditions were required to be fulfilled for such experiments—the conditions depending on the perceived danger of each experiment. These regulations also called for the development of "safer" vectors. Second, improved vectors were required because very large numbers of recombinant clones are needed to ensure that all possible genomic sequences have been cloned. Calculations showed that with cloned DNA fragments 20 kb long, some 690,000 clones would be required for a 99% probability of finding a given sequence in a mammalian genomic library (Clarke and Carbon,

1976). Third, new methods were needed to identify clones containing the target sequence among all these clones.

The restrictive conditions for performing recombinant DNA experiments were most severe in the United States. Each experiment had to be assessed on the basis of the source of the DNA to be cloned and the type of vector to be used. Gradually, as experience accumulated, there came the realization that the dangers of most of these experiments had been overestimated and there came about a careful, but to many scientists a too slow, revision of the guidelines.

Conditions could be relaxed in part because of the development of new vectors that both were safer and more useful for gene cloners. These new vectors were based on bacteriophage λ rather than on plasmids (Murray and Murray, 1974). Bacteriophage vectors have the advantages that they can accept much larger fragments of DNA than plasmid vector. Nonessential genes from the center of the linear phage genome can be removed to make room for cloned DNA while the left and right "arms" of the phage DNA that contain the genes necessary for growth of the phage remain. These arms can be ligated to the foreign DNA and the recombinant DNA packaged in the phage coat proteins. Just as important, many hundreds of thousands of recombinant phage can be tested simultaneously by infecting bacterial cells growing as lawns on large agar plates. The new phage vectors were exemplified by the Charon series of bacteriophage vectors developed by Blattner and his colleagues (Blattner et al., 1977). Named, Charon, for the mythical boatman that ferries the souls of the dead across the Styx, these vectors can accommodate DNA fragments between 7000 and 22,000 nucleotides long. The Charon vectors were considered safe because they had been altered so that they could grow only in certain strains of bacteria that do not occur naturally. A further improvement came with the development of *in vitro* packaging, a test tube method for assembling infectious phage particles containing recombinant DNA molecules (Hohn and Murray, 1977; Sternberg et al., 1977). Using infectious phage particles is a very efficient way of introducing the recombinant DNA molecules into bacterial cells.

The final problem—that of detecting clones with the desired recombinant DNA molecules—was overcome by two developments of the technique devised by Southern[20] for detecting and sizing DNA fragments (see below). The techniques involved the transfer of bacterial colonies (Grunstein and Hogness, 1975) or bacteriophage plaques (Benton and Davis, 1977) to a nitrocellulose filter by simply laying the filter on the surface of the agar bearing the colonies or plaques. The phage and bacteria absorbed on the filter are lysed and hybridized with an appropriate radioactively labeled probe.

Once again, the globin genes provide an early example of the application of these new tools (Maniatis et al., 1978) and the general approach is still used. The first step is to isolate high molecular weight DNA from cells and to treat this with a restriction enzyme. DNA fragments 20 to 24 kb in size are isolated from the treated DNA and ligated to the left and right arms of the phage vector. These DNA molecules are packaged in phage coat proteins and the phage particles mixed with bacterial cells and plated on agar. Each phage infects a bacterial cell and the progeny of that phage infect surrounding cells. The agar becomes covered with clear plaques where phage have lysed the cells. One of the libraries prepared by Maniatis and colleagues was of rabbit DNA. They obtained 3.8×10^4 plaques per μg of DNA for a total of 780,000 phage clones. Of these phage, 97.4% contained rabbit DNA sequences and the average size of these rabbit sequences was 17 kb. A most important figure for genomic libraries is the estimate of how many recombinant clones are needed to ensure that there is a high probability that every sequence is present in the library. The rabbit genome was estimated to be 3×10^9 base pairs and for cloned fragments of 17 kb, a minimum of 810,000 clones would be needed for a probability of 99% that any given sequence is present in the library. These calculations indicated that Maniatis and co-workers had probably cloned the complete rabbit genome. The final test, of course, was to find interesting genes. A globin cDNA was used to screen all 750,000 plaques using the hybridization technique devised by Benton and Davis (1977). Four positive clones were found—identical to the expected number. These were further analyzed by restriction enzyme analysis (see below) and were found to contain β-globin sequences.

Contemporary Cloning

Cloning became an indispensable tool of molecular biology and genetics. There have been many subsequent improvements in the vectors used for cloning. These have been directed largely at finding vectors that are ever more convenient to use; that function in different host cells; that express cloned sequences efficiently; and that can accept ever larger pieces of DNA. Other plasmids were constructed that were more versatile. pBR322 became one of the most popular. Developed by Bolivar and his colleagues (Bolivar et al., 1977; Bolivar, 1988) from PMB9, itself a derivative of pSC101, pBR322 had two selectable markers and was smaller and thus more stable. It became the plasmid vector of choice when Sutcliffe determined its complete DNA sequence (see below). Hybrid vectors such as cosmids that combined attractive features of both plasmids and phage vectors were engineered (Collins and Hohn, 1978).

When the genome projects got underway in the late 1980s, new requirements had to be met. The key problem is that cloned DNA fragments have to be reassembled in the order that they occur in the chromosomes. This reassembly process is impossible with 700,000 cosmid clones—vectors that could accept very large pieces of DNA were needed. The key components of a yeast chromosome are its centromere; the telomeres at the ends; and a sequence that initiates replication of the chromosome, the autonomously replicating sequence. It was realized by Olson and his colleagues (Burke et al., 1987) that these elements could be assembled together with genomic DNA fragments and selectable markers to form so-called yeast artificial chromosomes (YACs). On introduction into yeast cells, YACs are replicated along with the bona fide yeast chromosomes. The special feature of YACs is that they can contain fragments of DNA as large as 1 megabase (1,000,000 base pairs) in length (Haldi et al., 1994). YACs are not without their problems. YAC libraries are more difficult to maintain; two or more DNA fragments are often cloned in the same YAC; and some YAC clones are unstable, losing their cloned DNA fragment. Nevertheless, YAC clones have proved invaluable as a means of structuring and organizing the myriad of smaller cosmid clones (see Nelson and Brownstein, 1994, for a review of YACs).

The successes, and the problems, of YACs have led to a variety of other artificial chromosome cloning vectors that will accept large fragments of DNA. BACs, for example, are bacterial artificial chromosomes that use the *E. coli* fertility plasmid, F-factor (Shizuya et al., 1992). They can accept inserts up to 350 kb, are stable and very efficient at transforming *E. coli*. PACs are based on a bacteriophage P1-based vector (Sternberg et al., 1977). These are selectable but accept only relatively small (<100 kb) and have to be packaged. Other very important vectors include those designed for expression of cloned genes and phage vectors such as M13 that are used extensively in DNA sequencing.

Making Synthetic DNA

Cloning is not the only way of making large amounts of a specific nucleic acid sequence. Indeed, the tools and techniques for making DNA and RNA predate cloning. Nucleic acid chemistry has a long and honorable pedigree as part of organic chemistry, from the initial characterization of the nucleotides through the work of Todd and his colleagues in the 1950s that showed how the nucleotides are linked together. It was as early as 1955 that Michelson and Todd linked together two thymidine molecules using a chemical strategy that formed the basis for all subsequent developments. The culmination of this early phase of polynucleotide synthesis was the assembly of the structural gene for an alanine transfer RNA by Khorana's group in 1972 (Khorana[13] et al., 1972; Khorana, 1979). It was estimated that this epic synthesis had required 20 man-years of work.

While this was a great achievement, the techniques used were not amenable to widespread implementation. This required continuing developments along two lines. One was the utilization of solid phase synthesis in which one end of the polymer (peptide or polynucleotide chain) is anchored while units (amino acids or nucleotides) are added to the free end. Known best in Merrifield's implementation for synthesizing peptides, solid phase synthesis has the same advantages for nucleic acid synthesis. Because the growing polynucleotide chain is attached firmly at one end to an inert support, it is easy to add components; permit them to react; wash away those components; add further reactants, and so on. The only manipulations required are the flow through of reagents or washes—manipulations that can be easily automated (Hunkapiller et al., 1984). Indeed, the development of automated machines was a critical step, enabling nonchemists to make oligonucleotides and use them for biological experiments.

The second line of development was to find better ways of ensuring that only the groups on the nucleotide involved in making the internucleotide linkage were available for reactions. This requires protecting the other chemical groups from degradation and reaction. Currently, tritylated nucleoside phosphoramidites are used (Caruthers, 1985). These are stable at high temperatures and they are added very efficiently to a growing polynucleotide chain, enabling long chains to be synthesized. The addition of a single nucleotide requires four chemical reactions with washings between each reaction. Each cycle of reactions and washings takes about 12 minutes. When the synthesis is complete, the protecting groups are removed and the chains freed from the solid support. While addition of nucleotides is very efficient, some proportion of chains in each cycle fail to add the nucleotide. Thus the final product (the longest chain) must be purified using high performance liquid chromatography.

While the initial impetus for making oligonucleotides was to synthesize genes (Khorana, 1979), this has been taken over by cloning—that is, by isolating and multiplying naturally occurring genes. Nevertheless, oligonucleotides play a role in almost every area of recombinant DNA technology (Itakura et al., 1984; Smith, 1994). They are used (1) in manipulating DNA during cloning, (2) as probes for detecting cloned gene sequences and in analyzing genes, (3) to produce defined mutations at specific sites so as to study the functions of genes, and (4) to direct DNA synthesis in the polymerase chain reaction.

Modern DNA synthesizers can make chains 100 nucleotides long overnight compared to Khorana's synthesis of the alanine tRNA gene that was 77 nucleotides long and required several years work by a large team. Oligonuclcotides are something of unsung heroes in molecular biology but the successes of chemists and engineers in devising DNA synthesizers deserve to be celebrated as great achievements of molecular biology.

Polymerase Chain Reaction

If cloning is a biological technique for making DNA and synthetic DNA chemistry an artificial approach, then the polymerase chain reaction (PCR) is a vigorous hybrid. Devised by Mullis,[16] PCR has two virtues—specificity and amplification (Mullis et al., 1986; Mullis, 1990). It is a two-step process in which oligonucleotides direct where synthesis begins while enzymes carry out the synthesis itself.

The specificity arises from the rather remarkable fact that DNA polymerase cannot synthesize a new DNA strand on an existing single-stranded template except where there is a short double-stranded stretch of DNA. The small fragment of DNA that binds to the single-strand to create this short double strand is called a primer. *In vivo*, during the replication DNA, these primers are short sections of RNA made by an enzyme called primase. *In vitro*, the synthetic oligonucleotides used are synthesized with known sequences such that they will hybridize to specified regions of a single-stranded DNA template. Calculations show that for a genome of the size and complexity of the human genome, an oligonucleotide as short as 19 nucleotides long will specify a unique sequence in that genome of 3×10^9 bases (Sambrook et al., 1989). Thus if a sequence is known, an oligonucleotide complementary to that sequence can be synthesized. DNA is heated so that the strands fall apart and the oligonucleotide hybridizes to its complementary sequence forming a double-stranded starting point for DNA polymerase to begin making a new DNA strand. A researcher can specify what DNA the polymerase will make by choosing the appropriate oligonucleotide primer.

Two primers that hybridize to sequences flanking the region of the target DNA to be synthesized are made for each reaction. One primer hybridizes to one strand and the other primer to the other strand. Polymerases attach and begin synthesizing new DNA strands using the existing strand as a template. Eventually, the polymerase falls off but not before the newly made strands extend beyond the position of the primer on the other strand. The reaction mixture is heated to separate the strands, the primers reattach and synthesis begins again. Products of this reaction cycle include segments of single stranded DNA corresponding exactly to the target sequence, bearing primer binding sites at each end. On subsequent cycles, these target sequences double in number so that approximately 2^{32} target sequences are made in 32 cycles. This is a vast number of molecules and PCR can be used to synthesize DNA from a single sperm that contains a single double-stranded DNA molecule.

In the original technique, DNA polymerase had to be added on each cycle because the enzyme was destroyed by the heating necessary to separate the DNA strands. A major advance came with the availability of a DNA polymerase isolated from the thermophilic bacterium *Thermus aquaticus* (Kaledin et al., 1980). This organism lives in hot springs at a temperature of 75°C and its polymerase will remain active through many cycles of PCR (Saiki et al., 1988). A second technological advance was the development of "thermal cyclers," automated machines that can be programmed to carry out the cycles of cooling and heating.

These technical advances together with the versatility of PCR has led to it being used in thousands of laboratories in all studies in which DNA is manipulated (Mullis et al., 1994). It is used to screen for recombinant clones by using as primers sequences present in the vector DNA flanking the cloning site; it is used to make DNA for sequencing; it is used for assembling clones by testing which clones carry common sequences; it is used in DNA-based diagnostics; it is used in ecology studies to determine which members of a population are genetically related; and, of course, it is used in forensics.

ANALYZING DNA AND GENES

No matter how pure a gene or DNA sequence and no matter in what quantities it is available, nothing much can be learned without the means of analyzing it. The advances I shall discuss in this section are analytical techniques that range from studying the large scale structure of genomes to determine the sequence, nucleotide by nucleotide, of genes, and, over the past few years, the nucleotide sequence of entire genomes.

Restriction Enzyme Mapping

Prior to cloning, other means had to be found to study eukaryotic genes. One approach was to exploit viruses such as adenovirus and SV40. These DNA tumor viruses bring about profound changes in the cells that they infect and so it was expected that their genes, having to interact with the machinery of mammalian cells, would have features similar to those of eukaryotic genes. Furthermore, these viruses have small genomes (5000 bases for SV40 and 36,000 bases for adenovirus) and thus gave the best hope for identifying genes, albeit viral, that cause cancer. By the late 1960s, genes had been found that acted immediately on infection of cells ("early" genes) while others were activated late in infection. A prerequisite to more refined analyses of these genomes was a map of where the genes are. In 1969 while on sabbatical in Israel, Nathans[24] learned of the restriction enzymes being isolated and studied by his colleague Hamilton Smith. Nathans realized that these enzymes might cut the SV40 genome reproducibly into smaller fragments that could be the basis for a gene map of the virus (Nathans, 1979). Danna and Nathans (1971) found that the enzyme cut SV40 DNA into seven pieces indicating that the circular SV40 DNA had seven sites that were attacked by this enzyme. The pieces were of differing sizes and were separated from each other by electrophoresis in polyacrylamide gels. Nathans further realized that the key to assembling these fragments in the correct order was to use other enzymes that cut at different sites and to use "partial digests" in which not every site available to an enzyme was cut by it. By 1973, Danna and Nathans had prepared the first restriction map, a detailed map of SV40.

The Danna–Nathans procedure soon underwent modifications. They detected fragments of SV40 DNA by growing the virus in cells cultured in medium containing radioactively labeled thymidine or phosphorous so that the viral DNA was radioactively labeled, followed by autoradiography of the polyacrylamide gels to reveal where the radioactive DNA fragments were located. This was not ideal. The autoradiographs took time to be exposed; polyacrylamide gels are inconvenient to prepare; and it is always best to avoid using radioactive materials if possible. A simple but very effective modification was made by Sharp et al. (1973) when they used ethidium bromide to stain the DNA fragments in the gels. Ethidium bromide molecules intercalate between the bases of a DNA molecule and when illuminated with UV light, the DNA fluoresces bright purple. This is very sensitive, capable of detecting as little as 0.05 μg of DNA. And a nontechnical but crucial development was the production of restriction (and other) enzymes by companies that relieved researchers of the chore of making their own enzymes.

These improvements are used to the present day, but while gels allow determination of the sizes of the fragments, they provide no information about what genes are present on the fragments. Furthermore, this approach does not work with more complex genomes because so many DNA fragments are produced. One of the most famous techniques in molecular genetics—Southern blotting devised by Ed Southern—solved this problem (Southern, 1975). This does for DNA electrophoresis what the Grunstein and Hogness technique (published the same year) did for cloning—that is, marrying sequence detection by hybridization with another technique. The essence of Southern blotting is simplicity itself. Following electrophoretic separation of DNA fragments in an agarose gel, the fragments are transferred from the gel to a solid support, a sheet of nitrocellulose of the kind used for ultrafiltration. Transfer is carried out by placing the gel on a sheet of filter paper that acts as a wick, drawing up buffer from a reservoir. The nitrocellulose sheet is laid on top of the gel and followed by a second sheet of filter paper. Finally, absorbent material such as paper towels are placed on top (Figure 1). Buffer is drawn from the reservoir through the gel and carries DNA from the gel through the nitrocellulose. The nitrocellulose carries a positive charge so the negatively charged DNA pieces are retained on the filter which is further treated to fix the DNA to it irreversibly. Hybridization probes can be added and washed off under conditions that prevent nonspecific binding of the probe and, following drying, the filter is exposed to X-ray film to reveal the sites of hybridization. In this way, a specific sequence can be detected among all the fragments derived from a restriction enzyme digest of a human genome.

It was not long before the procedure became known as Southern blotting, to be followed by Northern blotting (for transfer of RNA) and Western blotting (for transfer of protein). As to the origin of Southern blotting, it appears that Southern's school teacher may have played a hitherto unappreciated role in the development of molecular biology. Southern recalls that he was given the task of reproducing examination papers. In Southern's school, in days before "xerox" became a verb, this was done by writing in ink on a glossy sheet of paper. The writing was absorbed by a gelatin block and this block was used to transfer copies to paper! (personal communication).

Pulsed-Field Gel Electrophoresis

By the early 1980s it was apparent that there would be great gains in cloning and mapping if much larger pieces of DNA could be manipulated. But, how could these be analyzed? Small fragments are well resolved on polyacrylamide gels while agarose gel electrophoresis is an excellent method for separating DNA fragments that are 300 to 25,000 nucleotides long. Above that limit, DNA fragments migrate very slowly into the agarose and very large DNA molecules do not enter the gel at all. The solution came with the development of a novel and unexpected form of gel electrophoresis. Schwartz and Cantor (1984) found that very large DNA molecules could be forced to migrate through low density (1.5%) agarose gels by having two electrical fields set perpendicular to each other and applying them alternately for

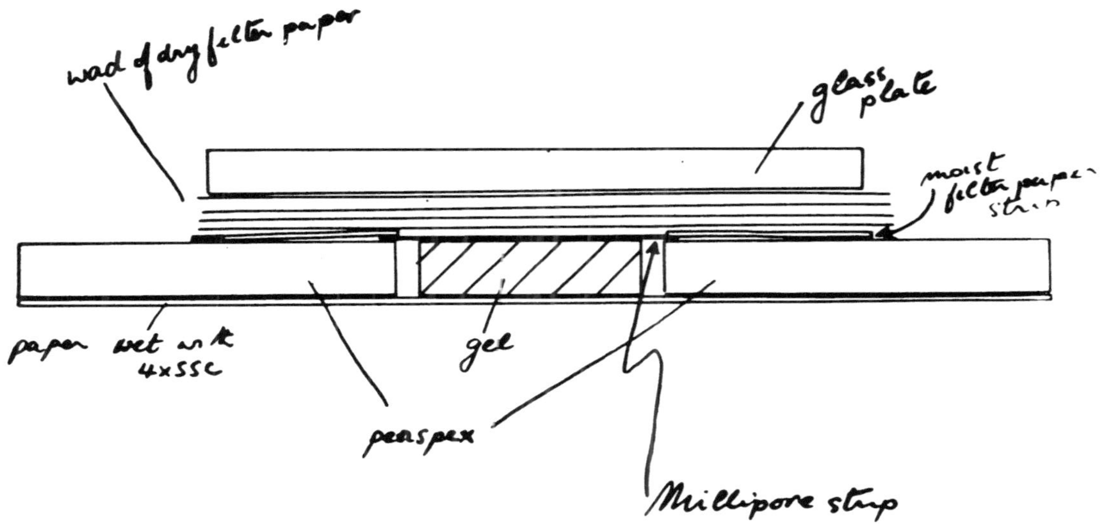

Figure 1. Southern's original drawing of the "blot" process.

short periods (hence "pulsed field" electrophoresis). For example, 11 yeast chromosomes could be resolved using two fields of differing strengths applied alternately for 45 seconds. These included the smallest (chromosome 3–300 kb) and largest (chromosome 12–1700 kb) chromosomes. An interesting problem was to prepare very large intact DNA molecules. It was clear that routine manipulations in molecular genetic laboratories were inadequate—simply drawing a solution of DNA into a pipette tip sheared the molecules. Instead Schwartz and Cantor embedded yeast cells in small agarose blocks and split the cells open and carried out other manipulations in the gel blocks. These were inserted into the gel slab for electrophoresis.

Early pulsed field gel electrophoresis was carried out with homemade apparatus and it was by no means a routine technique, requiring considerable application to determine the optimal conditions for separating large molecules. However, many modifications and variations of the basic principle were tried and as these were established and commercial apparatus became available, PFGE did become routine. PFGE in conjunction with YAC cloning and restriction mapping is an indispensable tool for laboratories carrying out genomic analysis and mapping.

DNA Sequence Determination

Even before the decipherment of the genetic code, determining the sequence of nucleotides in a DNA molecule was a challenge to biochemists and organic chemists. But a direct chemical attack on this problem seemed impossible and indirect methods, trying to deduce sequence from analyses of bulk DNA, were all that was available. It was a discouraging task. Chargaff,[4] DNA chemist *par excellence*, concluded in a review published in 1961: "In considering the problem of the nucleotide sequence in deoxyribonucleic acids we have barely turned the corner. There is a long road before us; and we shall not see its end." It was a long road but not as long as Chargaff predicted. Fifteen years later its end was reached.

Any discussions of sequence must begin with Sanger[12] whose autobiographical chapter for *Annual Review of Biochemistry* (Sanger, 1988) is entitled appropriately "Sequences, Sequences, Sequences." The first sequences are those of proteins; the second, RNA; and the third, DNA. Sanger's method for determining amino acid sequence depended on using proteolytic enzymes to break the polypeptide chain into smaller fragments that were separated by paper chromatography. This breaking-down phase was followed by a synthetic phase when the data were reassembled to yield the complete sequence. This same approach was used by Holley[10] and his colleagues (Holley, 1966) to sequence alanine tRNA, using ribonuclease T_1 (which cuts at G residues) to degrade the RNA and ion exchange column chromatography to separate the degradation products. This was the first RNA and nucleic acid to be sequenced. Sanger's improvement was to use two-dimensional chromatography on cellulose acetate and ion exchange paper rather than column chromatography. Using this and other techniques, in 1968 Brownlee, Sanger, and Barrell were able to report the sequence of 5S ribosomal RNA with 120 nucleotides.

Unfortunately, the same approach could not be used with DNA because enzymes with the specificity of proteases for proteins or ribonuclease T_1 for RNA were not available in the late 1960s. Restriction enzymes were being developed at this time and while they came to play an important role in DNA sequencing, it was not because of their ability to break down DNA molecules. Instead methods based on DNA synthesis were developed. The potential of this approach was demonstrated by Wu and Taylor who were able to determine the sequence of the *cos* ends of λ bacteriophage (Wu and Taylor, 1971; Wu, 1994). These cohesive ends are single strands of DNA that enable the infecting bacteriophage DNA to circularize prior to packing in the bacteriophage head proteins. DNA polymerase used the ends of the double strand as a primer for initiating synthesis on the single strand and Wu and Taylor supplied only one, two, or three of the necessary nucleotides so that synthesis was only partial. Analysis of the products was time consuming requiring one year to determine the sequence of the 12 nucleotides of the *cos* ends.

Sanger's contributions were twofold (Sanger, 1981, 1988). The first was to ensure that chains initiated at the same primer and ending at successive bases away from the primer were present in equimolar concentrations (Sanger et al., 1977). The solution came from work in Kornberg's laboratory where dideoxynucleotides (specifically dideoxythymidine triphosphate, ddTTP) were shown to terminate DNA synthesis because they did not have the 3′ hydroxyl necessary for the incoming nucleotide to form a phosphodiester link. The second insight was that polyacrylamide electrophoresis could be used to separate newly synthesized strands differing in length by only one base pair. Radioactive oligonucleotides 300 nucleotides long could be resolved on these gels and their positions detected by autoradiography with X-ray film.

Sanger and Coulson were given ddTTP by Geider and then synthesized the other three. Setting up four separate reactions containing one of each of the ddNTPs and the three remaining nucleotides as deoxynucleotides, DNA fragments were produced in each reaction that terminated wherever the ddNTP in that reaction had been incorporated. The products of each reaction were separated on a polyacrylamide gel and the gel exposed to X-ray film. The result was a series of bands in each lane from which the sequence could be read off.

There were two problems with this technique. The first was to prepare pure, single-stranded DNA as the substrate for the synthesis reaction, and the second was the need to make new primers for each different DNA to be sequenced. Both problems were solved by Messing and his colleagues (Messing and Vieira, 1982) who introduced M13, a single-stranded filamentous phage, as a cloning vector for DNA sequencing. Cloning in M13 provides a means to prepare pure DNA for sequencing; it is a single-stranded substrate for sequencing; and the primer is an oligonucleotide that hybridizes to M13 sequences immediately adjacent to the cloning site. This "universal" primer will prime synthesis of any DNA molecule cloned into M13.

At the same time that these developments were going on in Cambridge, England and Cambridge, Massachusetts, Maxam and Gilbert were devising a technique that did involve degradation of the DNA molecule, but by chemical rather than enzymatic means (Gilbert, 1981). Gilbert[22] had been interested in the control of gene expression and with Muller-Hill had isolated the *lac* repressor, the protein that regulates expression of the *lac* operon. He had followed that up by isolating the sequence to which the repressor binds by treating DNA first with the repressor and then deoxyribonuclease that destroyed all the DNA except the segment protected by the bound protein. (This technique became generalized as DNA "footprinting.") Gilbert and Maxam (1973) determined the sequence of this 25-nucleotide long fragment by converting it into RNA using reverse transcriptase and sequenced it using Sanger's RNA methods. Gilbert continued to isolate and characterize the *lac* operon region by sequencing mutations, but it was Dickson et al. (1975) who determined the sequence of the *lac* promoter.

Accordingly, wrote Gilbert some years later, ". . . by the middle 1970s I knew all the sequences I had been curious about . . ." (Gilbert, 1981). His interest in developing a new sequencing technique was suggested by the work of Mirzabekov who was using dimethyl sulphate to methylate adenines and guanines in DNA molecules. Gilbert realized that such methylated nucleotides could be removed by heat leaving only the sugar phosphate chain holding the strand together at those positions. These links were weak and could be easily broken by hydrolysis. The resulting fragments could be detected by electrophoresis and autoradiography if the 5′ end of the molecule had been labeled. In essence, one would get a set of fragments ending in an A or a G. It worked. It was now necessary to find a way to detect cytosines and thymidines. This was achieved by using hydrazine which, in the presence of high salt concentrations, reacts only with cytosines which can be cleaved using piperidine. In low salt concentrations, both cytosines and thymidines react with hydrazine. Sequencing using the Maxam–Gilbert (1975) method involves, then, setting up four reactions that give the positions of A, G, C, and C with T, the positions of the Ts being deduced. The power of this new sequencing method became abundantly clear when Sutcliffe sequenced the β-lactamase gene of the cloning plasmid pBR322; he learned the technique and obtained 1000 nucleotides of completed sequence in seven months (Sutcliffe, 1978). From there, Sutcliffe went on to determine the full 4363 base pairs of the pBR322 sequence (Sutcliffe, 1979, 1995).

Initially the Maxam–Gilbert strategy was probably preferred, but with the development of M13 as a vector for producing the single-stranded DNA needed for the dideoxy termination method, the latter has become the standard method. This is particularly true with the development of automated DNA sequencing machines that use fluorescently labeled primers to distinguish the products of the four termination reactions (Smith et al., 1986). The four sequencing reactions can be mixed and analyzed on a single lane because a laser detects each oligonucleotide as it moves down the gel by the specific emission of each dideoxy terminator. This has led to a very significant increase in the rate of sequencing. Modern machines have 36 lanes, each of which can resolve up to 550 nucleotides. In specialist laboratories, the machines are run two times each day and once on Saturdays and Sundays, for a weekly throughput of 237,600 nucleotides per machine. Thus in early 1995, the sequencing group at St. Louis was able to enter, over a five-month period, 101,458 sequences, averaging 357 nucleotides (for a total of 36 mb), into the dBEST database.

FUNCTIONAL ANALYSIS OF DNA

By 1975, it was possible to isolate genes and determine their sequence, but this was only the first part of understanding the genetic basis of biological phenomena. What was required was a molecular equivalent for the mutations studied in classical genetics, although the process would be reversed in the molecular biologist's laboratory. That is to say, in the "classical" approach, a mutant phenotype is observed (either naturally occurring or caused by chemical mutagens or radiation) and this leads to the gene. In molecular genetics, on the other hand, a gene can be cloned and subsequently mutated and introduced into cells or animals so that the function of the gene can be inferred. Tools to do this were developed over a period of some 17 years.

Mutagenesis

The availability of the cloned DNA sequences for genes provided the opportunity for engineering specified mutations in the genes and then testing the effects of those mutations. There are two basic strategies for performing mutagenesis of cloned DNA. The first is random mutagenesis where mutations are induced in the cloned DNA and then the mutagenized clones are tested for a specific function. The second form of mutagenesis *in vitro*—site-directed mutagenesis—introduces single nucleotide changes at specified positions and is used for fine mapping of genetic function.

The simplest forms of mutagenesis involve insertions and deletions of nucleotides because these can be produced easily using restriction and other enzymes. For example, Lai and Nathans (1974) prepared SV40 mutants by using two enzymes, *Eco*RI and *Hin*dIII, to remove a segment of the SV40 genome. A further development used restriction enzymes that cut the cloned DNA only once followed by S1 exonuclease treatment to remove the single-stranded ends. Conversely, nucleotides can be added at a restriction site by cutting with the enzyme and then using DNA polymerase to synthesize a double strand on the single-stranded ends produced by the enzyme. Both procedures create blunt ends that can be joined by DNA ligase. If *Eco*R1 is the restriction enzyme, then the effect is to remove or insert four nucleotides at the restriction enzyme site. These methods have drawbacks. The mutations are produced only at restriction enzyme sites and they are not point mutations.

Many other ingenious techniques were devised (Watson et al., 1992) but the most important advance came with the use of synthetic oligonucleotides (Smith, 1994). The first experiments were performed using the single-stranded phage, øX174. Hutchison and Smith realized that an oligonucleotide complementary to a sequence in øX174 but with a single base change would hybridize to the wild-type sequence and could serve as a primer for replication of the single strand by DNA polymerase (Hutchison et al., 1978). The product of this reaction is a circular, double-stranded DNA molecule with the mutant oligonucleotide in one strand. When these molecules are transfected into *E. coli*, phage containing either the wild-type strand or the mutated strand are obtained. Hutchison et al. used an oligonucleotide 12 nucleotides long that hybridized to nucleotides 582 to 593 and contained a single nucleotide change of G to A at position 587. This is the øX174 mutant, *am3*. The mutant phage had an A at position 587 and the expected phenotype.

øX174 is not a useful phage for recombinant DNA work because it cannot accept DNA inserts without exceeding the size that can be packaged. However, at the same time that these first oligonucleotide mutation strategies were being devised, significant advances were being made in using phage as vectors. In particular, the phage M13 was being used for sequencing (Messing and Veira, 1982). This became the standard vector for site-directed mutagenesis using oligonucleotides (Zoller and Smith, 1982). Subsequent improvements increased the efficiency of the procedure by using two primers, one mutagenic and the other a universal primer to increase the number of completed strands synthesized. Phage plaques are screened by hybridization with a probe that detects the mutated sequence. A further improvement came with the development of a technique that caused "self-destruction" of the nonmutated DNA strands (Kunkel, 1985). Phage are grown in *E. coli* that have the *dut*$^-$ and *ung*$^-$ mutations. The former leads to a deficiency of deoxyuridine triphosphatase so that intracellular dUTP accumulates and is incorporated into DNA in place of dTTP. The *ung*$^-$ mutation is in a gene for uracil-*N*-glycosylase, an enzyme that removes dUTPs that have been mis-incorporated so that phage grown in this *E. coli* double mutant accumulate as many as 30 uracils in their DNA. Following synthesis *in vitro* of the DNA strand containing the mutant oligonucleotide, the double-stranded molecules are transfected into *ung*$^+$ bacteria that have uracil-*N*-glycosylase that destroys the wild-type DNA strands that had incorporated uracil. As many as 80% of the phage derived from these cells have the mutated DNA.

Introducing Nucleic Acids into Cells

By the mid-1980s, the tools were available for producing mutations in any piece of DNA whose sequence was known and which could be cloned in M13. But these molecules then had to be introduced into cells to determine the biological consequences of the mutations. Methods for transferring naked DNA into bacterial cells were, of course, an integral part of cloning, but doing the same for mammalian cells was more difficult (Watson et al., 1992). The first methods to be devised used complexes of DNA with calcium phosphate (Graham and van der Eb, 1973) or DEAE-dextran (McCutchan and Pagano, 1968). For calcium phosphate transfection, DNA is simply mixed with a calcium phosphate solution and left for about 30 minutes during which time precipitates form. These are added to mammalian cells in tissue culture where they are taken up, presumably by endocytosis. In some way, the DNA is transferred to the nucleus where it becomes stably integrated into cell chromosomes. Usually multiple copies of the DNA integrate. Not all cells are amenable to calcium phosphate transfection but will take up DNA complexed with DEAE-dextran which has the added advantage of needing less DNA.

The most significant use of DNA transfection into mammalian cells came with studies in the late 1970s and early 1980s that cloned oncogenes using these techniques (Der et al., 1982; Goldfarb et al., 1982; Shih and Weinberg, 1982). DNA was extracted from tumors or from tumor cells growing in tissue culture and transfected into NIH/3T3 cells. After several weeks, groups of cells were found that were growing faster than their neighbors and were piling up. These cells were isolated and their DNA extracted and added to fresh NIH/3T3 cells. Some of these in turn were transformed, indicating that a stable DNA sequence was being transmitted. This sequence–an oncogene–could be cloned using the human-specific *Alu* sequence as a tag or by hybridizing DNA from secondary transformants with probes from retroviral oncogenes.

There are three remarkable "physical" techniques for introducing DNA into cells. In one method, DNA is injected directly into the nuclei of cells growing in monolayer tissue culture (Capecchi, 1980). This method is very efficient and very slow, although computer-guided apparatus now relieves a scientist of the tedious task of sitting for hours at the microscope. Electroporation uses pulses of high voltage (between 250 V/cm and 750 V/cm) to make holes in cell membranes through which DNA passes (Neumann et al., 1982). As many as 80% of the cells may be killed, but very large numbers of cells can be treated in suspension so that many survive. Plant cells are particularly difficult targets because of their cell walls. Microbombardment uses small (1–2 μM diameter) beads of tungsten or gold coated with the DNA to be transfected (Klein et al., 1987). These are "shot" into the plant tissue where they break through the cell walls. This technique has been adapted for animal cells and tissues.

Finally, biological methods have been used, notably vectors based on retroviruses and other viruses that infect animal cells. These have the tremendous advantage that they have evolved so that they are extremely efficient at transferring their own DNA into cells. Geneticists can exploit this by cloning genes into "stripped-down" versions of retroviruses that can infect cells but not replicate. For example, the *gag*, *env*, and *pol* genes of a retrovirus are essential for replication of the virus but are not necessary for infection. Thus these genes can be removed and replaced by a cloned gene for introduction into the cells. This vector retains the long terminal repeats needed for integration into the host cell DNA and the Ψ-sequence needed for packaging the viral genomic RNA. Infectious vector virus has to be made by transfecting this DNA into a packaging cell line that contains an integrated form of the virus that lacks the Ψ-sequence but has the *gag*, *env*, and *pol* genes. This produces the proteins needed for packaging the vector RNA, but its own genomic RNA cannot be packaged without the Ψ-sequence (Mann et al., 1983). Other viruses being adapted for use as vectors include adenovirus which can accept up to 7 kb of cloned gene. Adenovirus is particularly attractive for developing gene therapy for cystic fibrosis because the virus has evolved to infect epithelial cells.

These vectors have been developed especially for use in gene therapy where cells are removed from a patient; the appropriate gene is introduced via a retroviral vector; and the cells are grown in culture and then put back into the patient. Hypercholesterolemia (Grossman et al., 1994) and adenosine deaminase deficiency have been treated using retroviral vectors. Patients with cystic fibrosis have received adenovirus with the cystic fibrosis transmembrane conductance protein by aerosol spray directly into the lungs.

Targeting Genes

These methods of introducing genes into cells work well except that the experimenter has neither control over the numbers of copies of the transgene that become integrated nor over the site of integration. As well as simply adding genes to a cell, it would be very useful to be able to knock out a gene *in vivo* so as to observe its functions, or even to introduce defined mutations *in vivo* (Sedivy and Joyner, 1992). The technique for doing this was obvious from work in bacteria and yeast—homologous recombination. In these organisms, introduced genes recombine with their homologues *in vivo* at high frequencies. Unfortunately, DNA introduced into mammalian cells almost invariably undergoes nonhomologous recombination and becomes integrated at random. It is not known why mammalian cells possess such high rates of nonhomologous recombination; it has been suggested that it is a protective device that ensures that any chromosomal fragments are retained by the cell, albeit in an inappropriate place. While this may be useful to the cell, it is the bane of scientists trying to target transgenes.

Nevertheless, the potential of this approach was demonstrated by Smithies et al. (1985) who showed that it was possible to target homologous recombination using the β-globin gene with a frequency of about 1 in 1000 transformed cells. Further developments primarily by Capecchi (Thomas and Capecchi, 1987; Mansour et al., 1988) and Smithies (Doetschman et al., 1988) were directed to devising powerful methods for selecting the rare homologous recombinants or sensitive detection systems. The former is exemplified by the "positive–negative" section technique. The transgene itself is disrupted by cloning into it an *aph* gene conferring resistance to a cytotoxic drug, G418, while a *tk* gene for thymidine kinase is cloned into the vector, downstream of the disrupted transgene. Cells producing thymidine kinase are sensitive to gancyclovir. Following exposure of cells to the vector, cells that have integrated the transgene can be selected for by using G418; this kills cells without *aph* (which comes with the transgene). Of these cells, those that have not undergone homologous recombination are killed by treating with gancyclovir because these cells have a *tk* gene (which is lost if the transgene has been incorporated by homologous recombination). An alternative strategy uses microinjection to ensure that cells receive the disrupted transgene. These cells are grown up and PCR is used to locate the clones in which the transgene has integrated by homologous recombination. Gene targeting has become a very powerful tool for analyzing the functions of cloned genes by mutating a gene in mice and examining the phenotype that results (Capecchi, 1989). Some examples are given in the next section.

Introducing Genes into Embryos

Transfection into cells in culture is sufficient if one is interested in only, for example, the regulation of gene expression. Something more is needed for studying the role of genes in development—the genes have to be introduced into embryos. The first experiments were performed by Brinster who microinjected DNA directly into the male pronucleus of a fertilized egg (Hammer, 1988). The eggs are transplanted to a pseudopregnant female and the offspring examined for expression of the transgene. Transgenic mice are bred to determine which have integrated the transgene in their germ cells. Of these early successes, the most spectacular were transgenic mice containing and expressing multiple copies of rat growth hormone gene (Palmiter et al., 1982). The mice were twice the body weight of their normal litter mates.

The real power of transgenic animals came with the development of embryonic stem cell culture and its combination with homologous recombination. Embryonic stem (ES) cells are derived from the inner cell mass, the part of the blastocyst that will become the embryo. ES cells can be grown in culture and they retain the ability to differentiate into all the cells types of the adult (Evans and Kaufman, 1981; Martin, 1981). If ES cells are injected into the cavity of a blastocyst, they will become part of the embryo and integrate into all adult tissues. The great advantage of ES cells over injection into fertilized eggs is that ES cells can be manipulated in culture. Thus cells carrying a transgene can be selected *before* their introduction into an embryo. This means that genes can be mutated in ES cells by homologous recombination, cells with the mutation selected by positive–negative selection or detected by PCR, and then introduced into blastocysts.

This is a key strategy in generating mouse models for human-inherited disorders where the gene involved has been cloned (Clarke, 1994). For example, homologous recombination has been used to produce mouse "knock-outs" with mutations in the genes involved in cystic fibrosis, the Lesch–Nyhan syndrome, and β-thalassemia.

In addition to these single-gene disorders, knock-outs have been made of many oncogenes, including *bcl-2*, *fos*, *src*, *jun*, and *myc*, and tumor suppressor genes such as *p53* and *rb*. Furthermore, the interactions between oncogenes is being studied by breeding strains carrying mutations in these genes. The importance of these mutant mice is such that an Induced Mutant Resource has been established at the Jackson Laboratory for maintaining stocks of these mice. The Resource contains over 130 mutations produced by gene targeting.

FROM COTTAGE INDUSTRY

Advances in the fields of molecular genetics and biology have depended not only on intellectual and technical advances. In addition, the development of what can be best described as commercial support services and ready access to methods that had been developed in pioneering laboratories are critical factors not often recognized.

Suppliers of Reagents

The impact of the development of a service industry can be seen best by examining the methods sections of papers published in the early days of recombinant DNA. Take, for example, the *Cell* paper by Maniatis et al. (1976) describing the cDNA cloning of rabbit β-globin. Of the reagents peculiar to recombinant DNA, the majority were either prepared in the Maniatis' laboratory or were gifts from colleagues (Table 1). Similarly, Sutcliffe (1978, 1995) describes how he characterized two new restriction enzymes in the course of his project sequencing pBR322. The Gilbert laboratory also prepared its own γ-labeled nucleotides for kinase labeling fragments; this involved using 25 to 50 mCi of radioactivity at a time!

A nonsystematic sampling of papers published in *Cell* in 1994 shows that none of the reagents is made by the laboratories describing the fruits of their research. Instead, these reagents are bought from commercial suppliers and this is now such a wide standard practice that the companies themselves are not named. Even cDNA libraries and Northern blots prepared from specific developmental stages and tissues can be purchased and companies will prepare cDNA libraries from mRNA supplied to them. While many laboratories synthesize their own oligonucleotides, large scale genomic mapping requires the preparation of very large numbers of oligonucleotides as probes. Companies will synthesize thousands of such custom made sequences.

The nine restriction enzymes listed in Table 1 are particularly noteworthy as restriction enzymes exemplify this move from cottage industry to commercial support. Ham Smith had recognized the usefulness of the enzymes he was studying but failed to interest companies in producing them. With Smith's encouragement, Donald Comb began making Hind II and Hind III and selling them to Miles Laboratories, although the process was slow because the assay (depending on

Table 1. Sources of Biological Reagents used in Maniatis et al. (1976)

Maniatis Laboratory	*Commercial Suppliers*	*Gifts*	*Source*
T4 polynucleotide kinase	DNA polymerase I	AMV reverse transcriptase	T. Papas
λ-exonuclease	DNAase I	*E. coli* DNA polymerase	W.R. McLure
	RNAase A	Exonuclease III	R.J. Roberts
		Restriction enzyme:BglI	R.J. Roberts & P. Myers
		Restriction enzyme:Bal	R.J. Roberts & P. Myers
		Restriction enzyme:Bam	R.J. Roberts & P. Myers
		Restriction enzyme:MboII	R.J. Roberts & P. Myers
		Restriction enzyme:Hph	R.J. Roberts & P. Myers
		Restriction enzyme:HaeIII	R.J. Roberts & P. Myers
		Restriction enzyme:AluI	R.J. Roberts & P. Myers
		Restriction enzyme:Hinf	R.J. Roberts & P. Myers
		Restriction enzyme:EcoRI	R.J. Roberts & P. Myers
		Terminal transferase	R. Ratcliff
		E. coli RNA polymerase I	B. Meyers
		Calf thymus RNAase H	J.G. Stavrianopoulos
		E. coli K12, HB101	H. Boyer
		E. coli SG5519	S. Gottesman

measuring changes in DNA viscosity) was imprecise. At about the same time, Rich Roberts at Cold Spring Harbor Laboratory started a systematic search and cataloging of restriction enzymes, and began publishing lists of these enzymes and their characteristics (see, for example, Roberts, 1980; Roberts and Macelis, 1994). In so doing his laboratory became a clearing house for restriction enzymes. Roberts met Comb and suggested a new assay for restriction enzymes, using the agarose gel electrophoresis and ethidium bromide staining developed at Cold Spring Harbor Laboratory. This revolutionized the process and restriction enzymes were prepared from many strains supplied by Roberts. The response of molecular biologists was very positive despite having to pay for what had previously been "free." The advantages of quality controlled reagents and kits and the time gained in not preparing them were recognized quickly and today no one would contemplate making their own enzymes. (Recently, the commercial supply of enzymes has been controversial over the use of *Taq* DNA polymerase. The patent on this enzyme is owned by Hoffman-La Roche which is vigorously enforcing its monopoly.)

Manuals

By the mid-1970s, it was clear that recombinant DNA techniques were essential not only for studying genes, gene structure, and function, but were also key tools for scientists studying many aspects of biology. However, these techniques had been developed in a relatively small number of laboratories and while papers published in this period gave details of the methods, the descriptions were characteristic of the terse style in which scientific papers are supposed to be written. In 1980, Cold Spring Harbor Laboratory held a laboratory course taught by Tom Maniatis, Ed Fritsch, and Joe Sambrook on the *Molecular Cloning of Eukaryotic Genes*. The manual for this course, assembled from procedures ". . . scattered throughout the notebooks of many different people," was circulated for comment and then published as *Molecular Cloning: A Laboratory Manual* (Maniatis et al., 1982). It was this book, with its mixture of recipes and basic knowledge underpinning the recipes, that gave many scientists the information and encouragement to acquire what had seemed to be the almost mystical skills of a craft. It is a measure of the rapid progress in the field of recombinant DNA, and an indication of the numbers of techniques molecular biologists are expected to master that *Cloning* tripled in size from 545 pages in 1982 to over 1400 pages in the second edition of 1989 (Sambrook et al., 1989). Since then there have been other manuals for molecular genetics but none with the impact of the first edition of "Maniatis." Thus Gilbert (1991) was able to write in *Nature* that "One looks up a recipe in the Maniatis book . . ." without having to provide a more detailed citation.

CONCLUSIONS

What of the future? I have not discussed the development of genomic research as a topic in its own right but I believe that we are about to enter a new phase in the study of biology as the complete genomic DNA sequences of organisms become available. The genomic sequence of *Hemophilus influenzae* was announced in May, 1995—the first genome of a free-living organism to be sequenced in its entirety. Several other genomes are approaching completion. As of June, 1995, some 77% of the sequence of *S. cerevisiae* is complete and it is likely that the entire sequence will be completed by the spring of 1996. Approximately 20% of the *C. elegans* sequence is complete and that should be completed in its entirety by 1998. Predictions for completion of the genomic sequence of *H. sapiens* have more uncertainty but it may be completed at a reasonable accuracy by early in the next century.

In 1991, Gilbert predicted that having genomic sequence available would change the way experimental biology is performed. A scientist interested in a particular biological phenomenon will first form some testable idea and then turn to the DNA databases to find sequences that can be used as tools to elucidate what is going on. These sequences could be used as nucleic acid probes, to produce a protein in bacteria for making monoclonal antibodies, or as oligonucleotides for site-directed mutagenesis. That this is becoming a reality is evident from the genomic studies of *S. cerevisiae*. Take, for example, the new genes discovered from analysis of the sequence. The three classes of new genes and the percentages they comprise are: (1) new genes, similar to known genes of known function (55%); (2) new genes, similar to known genes of unknown function (15%); and (3) new genes, unrelated to any known gene and of unknown function (30%). The functions of these latter genes is being explored experimentally by mutating them and determining changes in phenotype.

Just 28 years ago, Gunther Stent demonstrated the difficulties of making predictions when he wrote an article entitled, "That Was Molecular Biology, That Was" (Stent, 1968). In it, Stent suggested that by 1963 this research had entered an academic phase and that: "All hope that paradoxes would turn up in the study of heredity had been abandoned long ago, and what remained now was the need to iron out the details." He could not foresee the remarkable findings that were to flow from the recombinant DNA revolution. This revolution has been as profound in changing our knowledge and understanding of the living world as was the revolution brought about by the microscope some four centuries ago. Just as the microscope enlarged the study of living organisms by revealing cells as objects for study, so the tools of recombinant DNA have laid open new studies of living organisms through the analysis and manipulation of genes.

ACKNOWLEDGMENTS

I want to thank David Lipman, Wojtek Makalowski, Stephen Oliver, Bob Waterston, and Rick Wilson, who very kindly provided me with the data quoted in the section on the genome projects, and Nathaniel Comfort for his comments on the manuscript.

REFERENCES

Arber, W. (1979). Promotion and limitation of genetic exchange. Science 205, 361–365.

Arber, W. & Linn, S. (1969). DNA modification and restriction. Ann. Rev. Biochem. 38, 467–500.

Baltimore, D. (1970). Viral RNA-dependent DNA polymerase. Nature 226, 1209–1211.

Benton, W.D. & Davis, R.W. (1977). Screening λgt recombinant clones by hybridization to single plaques *in situ*. Science 196, 180–182.

Berg, P. (1981). Dissections and reconstructions of genes and chromosomes. Science 213, 296–303.

Berget, S.M., Moore, C., & Sharp, P.A. (1977). Spliced segments at the 5′ terminus of adenovirus 2 late mRNA. Proc. Natl. Acad. Sci. USA 74, 3171–3175.

Blattner, F.R. et al. (1977). Charon phages: Safer derivatives of bacteriophage lambda for DNA cloning. Science 196, 161–169.

Bolivar, F. (1988). Plasmid pBR322: The multipurpose cloning vector. Focus 10, 61–64.

Bolivar, F., Rodriguez, R.L., Greene, P.J., Betlach, M.C., Heynecker, H.L., & Boyer, H.W. (1977). Construction and characterization of new cloning vehicles. II. A multipurpose cloning system. Gene 2, 95–113.

Brenner, S. (1988). Foreword. In: The Nematode *Caenorhabditis elegans* (Wood, W.B., Ed.), pp. ix–xiii. Cold Spring Harbor Laboratory Press, Cold Spring Harbor.

Brownlee, G.G., Sanger, F., & Barrell, B.G. (1968). The sequence of 5 S ribosomal ribonucleic acid. J. Mol. Biol. 34, 379–412.

Burke, D.T., Carle, J.F., & Olson, M.V. (1987). Cloning of large segments of exogenous DNA into yeast by means of artificial yeast chromosome vectors. Science 236, 806–812.

Capecchi (1980). High efficiency transformation by direct microinjection into cultured mammalian cells. Cell 22, 479–488.

Capecchi (1989). Altering the genome by homologous recombination. Science 244, 1288–1292.

Caruthers, M.H. (1985). Gene synthesis machines: DNA chemistry and its uses. Science 230, 281–285.

Chargaff, E. (1961). The problem of nucleotide sequence in deoxyribonucleic acids. In: Biological Structure and Function (Goodwin, T.W. & Lindberg, O., Eds.), Academic Press, London.

Chow, L.T., Gelinas, R.E., Broker, T.R., & Roberts, R.J. (1977). An amazing sequence arrangement at the 5′ ends of adenovirus 2 messenger RNA. Cell 12, 1–8. [This issue of Cell has three other papers from the Cold Spring Harbor Laboratory group.]

Clarke, A. (1994). Murine genetic models of human disease. Curr. Opinion Genet. Dev. 4, 453–460.

Clarke, L. & Carbon, J. (1976). A colony bank containing synthetic ColE1 hybrid plasmids representative of the entire *E. coli* genome. Cell 9, 91–99.

Cohen, S.N. (1988). DNA cloning: A personal perspective. Focus 10, 1–4.

Cohen, S.N., Chang, A.C.Y., Boyer, H.W., & Helling, R.B. (1973). Construction of biologically functional bacterial plasmids *in vitro*. Proc. Natl. Acad. Sci. USA 70, 3240–3244.

Collins, J. & Hohn, B. (1978). Cosmids: A type of plasmid gene-cloning vector that is packageable *in vitro* in bacteriophage λ heads. Proc. Natl. Acad. Sci. USA 75, 4242–4246.

Crick, F. (1988). What Mad Pursuit, pp. 144, 146. Basic Books, New York.

Danna, K. & Nathans, D. (1971). Specific cleavage of simian virus 40 DNA by restriction endonuclease of *Hemophilus influenzae*. Proc. Natl. Acad. Sci. USA 68, 2913–2917.

Danna, K., Sack, G.H., & Nathans, D. (1973). Studies of simian virus 40 DNA. VII A cleavage map of the SV40 genome. J. Mol. Biol. 78, 363–376.

Der, C.J., Krontiris, T.G., & Cooper, G.M. (1982). Transforming genes of human bladder and lung carcinoma cell lines are homologous to the ras genes of Harvey and Kirsten sarcoma viruses. Proc. Natl. Acad. Sci. USA 79, 3637–3640.

Dickson, R.C., Abelson, J., Barnes, W.M., & Reznikoff, W.S. (1975). Genetic regulation: The Lac control region. Science 187, 27–35.

Doetschman, T., Maeda, N., & Smithies, O. (1988). Targeted mutation of the *hprt* gene in mouse embryonic stem cells. Proc. Natl. Acad. Sci. USA 85, 8583–8587.

Efstratiadis, A., Kafatos, F.C., Maxam, A.M., & Maniatis, T. (1976). Enzymatic *in vitro* synthesis of globin genes. Cell 7, 279–288.

Evans, M.J. & Kaufman, M.H. (1981). Establishment in culture of pluripotential cells from mouse embryos. Nature 292, 154–156.

Gilbert, W. (1981). DNA sequencing and gene structure. Science 214, 1305–1312.

Gilbert, W. (1991). Towards a paradigm shift in biology. Nature 349, 99.

Gilbert, W. & Maxam, A. (1973). The nucleotide sequence of the lac operator. Proc. Natl. Acad. Sci. USA 70, 3581–3584.

Goldfarb, M., Shimizu, K., Perucho, M., & Wigler, M. (1982). Isolation and preliminary characterization of a human transforming gene from T24 bladder carcinoma cell. Nature 296, 404–409.

Graham, F. & van der Eb, A.J. (1973). A new technique for the assay of infectivity of human adenovirus 5 DNA. Virology 52, 456–467.

Gros, F., Hiatt, H., Gilbert, W., Kurland, C.G., Risebrough, R.W., & Watson, J.D. (1961). Unstable ribonucleic acid revealed by pulse labelling in *Escherichia coli*. Nature, Lond. 190, 576–581.

Grossman M., Raper, S.E., Kozarsky, K., Stein, E.A., Engelhardt, J.F., Muller, D., Lupien, P.J., & Wilson, J.M. (1994). Successful *ex vivo* gene therapy directed to liver in a patient with familial hypercholesterolemia. Nat. Genet. 6, 335–341.

Grunstein, M. & Hogness, D.S. (1975). Colony hybridization: A method for the isolation of cloned DNAs that contain a specific gene. Proc. Natl. Acad. Sci. USA 72, 3961–3965.

Haldi, M. et al. (1994). Large human YACS constructed in a rad52 strain show a reduced rate of chimerism. Genomics 24, 478–484.

Hammer, R.E. (1988). The scientific contributions of Ralph I. Brinster to understanding mammalian embryo development and eukaryotic gene expression. In: Cellular Factors in Development and Differentiation: Embryos, Teratocarcinomas, and Differentiated Tissues (Harris, S.E. & Mansson, P.-E., Eds.), pp. 1–30.

Hoagland, M. (1990). Toward the Habit of Truth. W.W. Norton, New York.

Hohn, B. & Murray, K. (1977). Packaging recombinant DNA molecules into bacteriophage particles *in vitro*. Proc. Natl. Acad. Sci. USA 74, 3259–3263.

Holley, R.W. (1966). The nucleotide sequence of a nucleic acid. Sci. Amer. 214, 30–39.

Hunkapiller, M. et al. (1984). A microchemical facility for the analysis and synthesis of genes and proteins. Nature 310, 105–111.

Hutchison, III, C.A., Phillips, S., Edgell, M.H., Gillam, S., Jahnke, P., & Smith, M. (1978). Mutagenesis at a specific position in a DNA sequence. J. Biol. Chem. 253, 6551–6560.

Itakura, K., Rossi, J.J., & Wallace, R.B. (1984). Synthesis and use of synthetic oligonucleotides. Ann. Rev. Biochem. 53, 323–356.

Jackson, D.A., Symons, R.H., & Berg, P. (1972). Biochemical method for inserting new genetic information into DNA of simian virus 40: Circular SV40 DNA molecules containing λ-phage genes and the galactose operon of *Escherichia coli*. Proc. Natl. Acad. Sci. USA 69, 3370–3374.

Judson, H.F. (1979). The Eighth Day of Creation. Jonathan Cape, London.

Kaledin, A.S., Slyusarenko, A.G., & Gorodetskii, S.I. (1980). Isolation and properties of DNA polymerase from extremely thermophilic bacterium *Thermus acquaticus* YT1. Biokhymia 45, 644–

Khorana, H.G. (1979). Total synthesis of a gene. Science 203, 614–625.

Khorana, H.G. et al. (1972). Studies on polynucleotides. CIII. Total synthesis of the structural gene for an alanine transfer ribonucleic acid from yeast. J. Mol. Biol. 72, 209–217.

Klein, T.M., Wolf, E.D., Wu, R., & Sanford, J.C. (1987). High-velocity microprojectiles for delivering nucleic acids into living cells. Nature 327, 70–73.

Kornberg, A. (1989). For Love of Enzymes, p. 159. Harvard University Press, Cambridge, Massachusetts.

Kunkel, T. (1985). Rapid and efficient site-specific mutagenesis without phenotypic selection. Proc. Natl. Acad. Sci. USA 82, 488–492.

Lai, C.J. & Nathans, D. (1974). Deletion mutants of SV40 generated by enzymatic excision of DNA segments from the viral genome. J. Mol. Biol. 89, 179–193.

Lobban, P.E. & Kaiser, A.D. (1973). Enzymatic end-to-end joining of DNA molecules. J. Mol. Biol. 78, 453–471.

Maniatis, T., Fritsch, E.F., & Sambrook, J. (1982). Molecular Cloning: A Laboratory Manual. Cold Spring Harbor Laboratory Press, Cold Spring Harbor, New York.

Maniatis, T., Hardison, R.C., Lacy, E., Lauer, J., O'Connell, C., Quon, D., Kee, S.G., & Efstratiadis, A. (1978). The isolation of structural genes from libraries of eucaryotic DNA. Cell 15, 687–701.

Maniatis, T., Kee, S.G., Efstratiadis, A., & Kafatos, F.C. (1976). Amplification and characterization of a β-globin gene synthesized *in vitro*. Cell 8, 163–182.

Mann, R., Mulligan, R.C., & Baltimore, D. (1983). Construction of a retrovirus packaging mutant and its use to produce helper-free defective retrovirus. Cell 33, 153–159.

Mansour, S.L., Thomas, K.R., & Capecchi, M.R. (1988). Disruption of the proto-oncogene *int-2* in mouse embryo-derived stem cells: A general strategy for targeting mutations to nonselectable cells. Nature 336, 348–352.

Martin, G.R. (1981). Isolation of a pluripotent cell line from early mouse embryos in medium conditioned by teratocarcinoma stem cells. Proc. Natl. Acad. Sci. USA 78, 7634–7638.

Maxam, A. & Gilbert, W. (1975). A new method of sequencing DNA. Proc. Natl. Acad. Sci. USA 74, 560–564.

McCutchan, J.H. & Pagano, J.S. (1968). Enhancement of the infectivity of simian virus 40 deoxyribonucleic acid with diethylaminoethyl-dextran. J. Natl. Cancer. Inst. 41, 351–357.

Mertz, J. & Davis, R.W. (1972). Cleavage of DNA by RI restriction endonuclease generates cohesive ends. Proc. Natl. Acad. Sci. USA 69, 3370–3374.

Messing, J. & Vieira, J. (1982). A new pair of M13 vectors for selecting either strand of double-digest restriction fragments. Gene 19, 269–276.

Mullis, K. et al. (1986). Specific enzymatic amplification of DNA *in vitro*: The polymerase chain reaction. Cold Spring Harb. Symp. Quant. Biol. 51, 263–273.

Mullis, K.B. (1990). The unusual origin of the polymerase chain reaction. Sci. Am. 262, 56–65.

Mullis, K.B., Ferre, F., & Gibbs, R.A., Eds. (1994). The Polymerase Chain Reaction. Birkhauser, Boston.

Murray, N.E. & Murray, K. (1974). Manipulations of restriction targets in λ-phage to form receptor chromosomes for DNA fragments. Nature 251, 476–481.

Nathans, D. (1979). Restriction endonucleases, simian virus 40, and the new genetics. Science 206, 903–909.

Nelson, D.L. & Brownstein, B.H., Eds. (1994). YAC Libraries. W.H. Freeman, New York.

Neumann, E., Schaefer-Ridder, M., Wang, Y., & Hofschneider, P.H. (1982). Gene transfer into mouse L-cells by electroporation in high electric fields. EMBO J. 1, 841–845.

Nomura, H., Hall, B.D., & Spiegelman, S. (1960). Characterization of RNA synthesized in *E. coli* after bacteriophage T2 infection. J. Mol. Biol. 2, 306–326.

Okayama, H. & Berg, P. (1982). High-efficiency cloning of full-length cDNA. Mol. Cell. Biol. 2, 161–170.

Palmiter, R.D., Brinster, R.L., Hammer, R.E., Trumbauer, M.E., & Rabbitts, T.H. (1976). Bacterial cloning of plasmids carrying copies of rabbit globin messenger RNA. Nature, Lond. 260, 221–225.

Rosenfeld, M.G., Birnberg, N.C., & Evans, R.M. (1982). Dramatic growth of mice that develop from eggs microinjected with metallothionein-growth hormone fusion genes. Nature 300, 611–615.

Roberts, R.J. (1980). Restriction and modification enzymes and their recognition sequences. Nucleic Acids Res. 8, r63–r80.

Roberts, R.J. & Macelis, D. (1994). REBASE—restriction enzymes and methylases. Nucleic Acids Res. 22, 3628–3639.

Rogers, M. (1977). Biohazard. Alfred A. Knopf, New York.

Rougeon, F., Kourilsky, P., & Mach, B. (1975). Insertion of a rabbit β-globin gene sequence into an *E. coli* plasmid. Nucleic Acids Res. 2, 2365–2378.

Saiki, R.K. et al. (1988). Primer-directed enzymatic amplification of DNA with a thermostable DNA polymerase. Science 239, 487–491.

Sambrook, J., Fritsch, E.F., & Maniatis, T. (1989). Molecular Cloning: A Laboratory Manual. Cold Spring Harbor Laboratory Press, Cold Spring Harbor, New York.

Sanger, F. (1981). Determination of nucleotide sequences in DNA. Science 214, 1205–1210.

Sanger, F. (1988). Sequences, sequences, and sequences. Ann. Rev. Biochem. 57, 1–28.

Sanger, F., Nicklen, S., & Coulson, A.R. (1977). DNA sequencing with chain-terminating inhibitors. Proc. Natl. Acad. Sci. USA 74, 5463–5467.

Schwartz, D.C. & Cantor, C.R. (1984). Separation of yeast chromosome-sized DNAs by pulsed field gradient gel electrophoresis. Cell 37, 67–75.

Sedivy, J.M. & Joyner, A.L. (1992). Gene Targeting. W.H. Freeman, New York.

Sharp, P.A., Sugden, B., & Sambrook, J. (1973). Detection of two restriction endonuclease activities in *Haemophilus parainfluenzae* using analytical agarose-ethidium bromide electrophoresis. Biochemistry 12, 3055–3063.

Shih, C. & Weinberg, R.A. (1982). Isolation of a transforming sequence from a bladder carcinoma cell line. Cell 29, 161–169.

Shizuya, H. et al. (1992). Cloning and stable maintenance of 300-kilobase-pair fragments of human DNA in *Escherichia coli* using F-factor-based vector. Proc. Natl. Acad. Sci. USA 89, 103–107.

Smith, H.O. (1979). Nucleotide sequence specificity of restriction endonucleases. Science 205, 455–462.

Smith, L.M. et al. (1986). Fluorescence detection in automated DNA sequence analysis. Nature 321, 674–679.

Smith, M. (1994). Synthetic DNA and biology. Bioscience Reports 14, 51–66.

Smithies, O., Gregg, R.G., Boggs, S.S., Koralewski, M.A., & Kucherlapati, R.S. (1985). Insertion of DNA sequences into the human chromosomal beta-globin locus by homologous recombination. Nature 317, 230–234.

Southern, E. (1975). Detection of specific sequences among DNA fragments separated by gel electrophoresis. J. Mol. Biol. 98, 503–517.

Stent, G. (1968). That was molecular biology, that was. Science 160, 395.

Sternberg, N. (1990). Bacteriophage P1 cloning system for the isolation, amplification, and recovery of DNA fragments as large as 100 kilobase pairs. Proc. Natl. Acad. Sci. USA 87, 103–107.

Sternberg, N., Tiemeier, D., & Enquist, L. (1977). *In vitro* packaging of a lambda Dam vector containing EcoRI DNA fragments of *Escherichia coli* and phage P1. Gene 1, 255–280.

Sutcliffe, J.G. (1978). Nucleotide sequence of the ampicillin resistance gene of *Escherichia coli* plasmid pBR322. Proc. Natl. Acad. Sci. USA 75, 3737–3741.

Sutcliffe, J.G. (1979). Complete nucleotide sequence of the *Escherichia coli* plasmid pBR322. Cold Spring Harbor Symp. Quant. Biol. 43, 77–90.

Sutcliffe, J.G. (1995). pBR322 and the advent of rapid DNA sequencing. Trends Biochem. Sci. 20, 87–90.

Temin, H.M. & Mizutani, S. (1970). Viral RNA-dependent DNA polymerase. Nature 226, 1211–1213.

Thomas, K.R. & Capecchi, M.R. (1987). Site-directed mutagenesis by gene targeting in mouse embryo-derived stem cells. Cell 51, 503–512.

Volkin, E. (1995). What was the message? Trends Biochem. Sci. 20, 206–209.

Volkin, E. & Astrachan, L. (1956). Intracellular distribution of labeled ribonucleic acid after phage infection of *Escherichia coli*. Virol. 2, 433–437.

Watson, J.D., Gilman, M., Witkowski, J., & Zoller, M. (1992). Recombinant DNA. W.H. Freeman, New York.

Watson, J.D., Hopkins, N.H., Roberts, J.W., Steitz, J.A., & Weiner, A.M. (1987). Molecular Biology of the Gene, Chapter 14. Benjamin/Cummings,

Watson, J.D. & Tooze, J. (1981). The DNA Story. W.H. Freeman, New York.

Winnacker, E.-L. (1987). From Genes to Clones. VCH Publishers, New York.

Witkowski, J.A. (1988). The discovery of 'split' genes: A scientific revolution. Trends Biochem. Sci. 13, 110–113.

Wood, W. B., Ed. (1988). The Nematode *Caenorhabditis elegans*. Cold Spring Harbor Laboratory Press, Cold Spring Harbor.

Wu, R. (1994). Development of the primer-extension approach: A key role in DNA sequencing. Trends Biochem. Sci. 19, 429–433.

Wu, R. & Taylor, E. (1971). Nucleotide sequence analysis of DNA. II. Complete nucleotide sequence of the cohesive ends of bacteriophage lambda DNA. J. Mol. Biol. 57, 491–511.

Ycas, M. & Vincent, W. (1960). A ribonucleic acid fraction from yeast related in composition to deoxyribonucleic acid. Proc. Natl. Acad. Sci. USA 46, 804–811.

Zoller, M.J. & Smith, M. (1982). Oligonucleotide-directed mutagenesis using M13-derived vectors: An efficient and general procedure for the production of point mutations in any fragment of DNA. Nucleic Acids Res. 10, 6487–6500.

Chapter 4

EXTRANUCLEAR DNA

Anil Day and Joanna Poulton

NON-MENDELIAN INHERITANCE

In the mid-nineteenth century Gregor Mendel's work on garden peas established rules that governed the inheritance of visible differences between parent plants. Mendel's work lay dormant for 40 years until it was rediscovered independently by Carl Correns, Hugo de Vries, and Erich Tschermark von Seysenegg in 1900. Mendel's rules stimulated much research on inheritance in the early twentieth century. This work established Mendel's laws on the segregation of alleles and the independent assortment of genes as the foundation of classical genetics. Work on meiosis, notably by Sutton in 1903, implicated chromosomes located in the nucleus as the agents responsible for Mendelian segregation.

While Mendel's legacy was being firmly established in the early part of the twentieth century there was little room for other models of inheritance. Apart from linkage, exceptions to Mendelism were largely overlooked. The influential American School of Geneticists, including Thomas Hunt Morgan and co-workers, concentrated on mapping chromosomes. Carl Correns studied the inheritance of white and pale green sectors on the leaves of variegated *Mirabilis jalapa* or four o'clock plants (1909). The white patches (chlorophyll-deficient areas) were only inherited by the progeny if they were present in the maternal parent. The male parent did not transmit the variegated phenotype. In other words, egg cells but not pollen transmitted the white trait. This maternal inheritance pattern was transmitted over many generations. This observation clearly contravened Mendelian inheritance where both parents transmit genes to their progeny. Erwin Baur (1909) studied the inheritance of variegation in *Pelargonium* and noted that both the male and female parts of flowers could transmit white leaves. However, the ratios of variegated to green progeny plants were not found in Mendelian proportions.

While it was clear that genes on chromosomes were responsible for Mendelian inheritance, the basis for non-Mendelian inheritance was unknown. Baur (1909) implicated the plastid family of organelles, whose best known member is the green chloroplast, as the physical entities responsible for non-Mendelian inheritance of green/white variegation in plants. Correns (1922) postulated a labile cytoplasm that could change into a normal (permanently green) or diseased (permanently white) state (reviewed in Kirk and Tilney-Basset, 1967). Hypothetical hereditary units in the cytoplasm called "plasmagenes" were also invoked to explain non-Mendelian inheritance (Lindegren, 1949). The existence of plasmagenes or extranuclear genes was supported by the work of Boris Ephrussi[34] (1949) on "petite colonie" mutations in bakers' yeast (*Saccharomyces cerevisiae*) and Ruth Sager's work (1954) on uniparental inheritance of antibiotic resistant mutations in the green alga *Chlamydomonas reinhardtii*.

In sexual crosses of bakers' yeast or *C. reinhardtii*, cells of opposite mating type fuse to form a diploid zygote that undergoes meiosis to form four haploid spores. *Petite* or small colonies of yeast grow very slowly because they cannot respire. Acridine dyes, that are now known to mutate DNA, promoted the production of *petite* mutations. Crosses of wild-type yeast cells with some classes of *petite* mutants show non-Mendelian segregation of the mutation producing either wild-type progeny exclusively or both wild-type and mutant in different proportions. In Mendelian inheritance each parent contributes chromosomal genes to two of the four haploid spores. The absence of respiratory enzymes associated with mitochondria in *petite* mutants implicated this organelle in the transmission of the *petite* state. Ephrussi postulated the presence of a heritable cytoplasmic unit responsible for the synthesis of respiratory enzymes in wild-type cells. This unit was absent in *petite* cells. In *C. reinhardtii* a mutation (*sr*-2) conferring resistance to streptomycin is inherited by all the progeny if it is present in the mating type plus parent. The mating type minus parent did not transmit streptomycin resistance to progeny cells. These results in *C. reinhardtii*, obtained in 1954, resembled the uniparental or maternal inheritance of white plastids published by Correns 45 years earlier.

THE SEARCH FOR EXTRANUCLEAR DNA: EARLY STUDIES ON ORGANELLE GENOMES

The discovery in the 1950s that genes were composed of DNA provided a firm basis for investigating cytoplasmic inheritance. Could DNA located in the cytoplasm account for the hypothetical plasmagenes responsible for cytoplasmic inheritance? Both plastids and mitochondria were known to divide from preexisting structures and could not be formed *de novo*. The work pioneered by Baur and Correns on green/white variegation in plants and Ephrussi on *petite* mutations in yeast suggested that extranuclear DNA may be located in plastids and mitochondria. Early attempts to detect DNA in plastids included the use of the Feulgen staining procedure which is specific for DNA, and also [^{3}H]thymidine to label DNA which could then be visualized by autoradiography. The results were equivocal (summarized in Kirk and Tilney-Basset, 1967). In 1962 electron microscopy of fixed cells treated with uranyl acetate revealed fibrils in the single cup-shaped plastid of *C. reinhardtii* (Ris and Plaut, 1962). These fibrils resembled the DNA fibrils present in bacteria and were removed by DNAse, an enzyme that specifically digests DNA. These observations provided convincing evidence for the presence of DNA in plastids. Subsequently, DNA fibrils were also detected in mitochondria (Nass and Nass, 1963).

The purification of mitochondria and plastids by differential centrifugation produced subcellular fractions that were enriched for these organelles. The job is made easier by choosing tissues that contain large numbers of the organelle being purified. Green leaves of young seedlings are particularly suitable for chloroplast isolation, while liver or heart were used for early preparations of animal mitochondria. Density gradient centrifugation provided a further step to obtain highly purified organelles. Bakers' yeast mitochondria band at a density of 1.16 to 1.18 g/cm^3 after centrifugation in linear 20–60% (W/V) sucrose gradients (Schatz, 1968). The availability of relatively pure preparations of intact chloroplasts or mitochondria allowed DNA associated with these organelle fractions to be analyzed. Contaminating DNA from the nucleus, released during homogenization, lies outside the organelles and can be removed by treating organelle fractions with DNAse. DNAase cannot enter intact organelles and only digests nuclear DNA contaminating the outsides of organelles and DNA present in broken organelles. Despite these precautions to obtain highly purified organelles, it was not sufficient to demonstrate that DNA could be detected in organelles. It was important to distinguish organelle-specific DNA from nuclear DNA using the analytical techniques that were available in the 1960s.

Base Composition and Conformation

The base composition, size, and conformation of organelle DNA were the main criteria used to distinguish it from nuclear DNA. The buoyant density of DNA is a reflection of its base composition. On caesium chloride or caesium sulphate density gradients, A+T-rich DNA has a lower buoyant density than G+C-rich DNA. In *C. reinhardtii*, plastid DNA has a buoyant density of 1.696 g/cm^3 and is relatively easily separated from nuclear DNA (1.723 g/cm^3) (Sager and Ishida, 1963). Chicken mitochondrial DNA (1.707 g/cm^3) bands below nuclear DNA (1.698 g/cm^3) but is too close to allow easy purification (Borst and Rutenberg, 1966). In higher plants the density of chloroplast DNA is also too close to that of nuclear DNA to allow their separation. In higher plants, the abundance of 5-methylcytosine

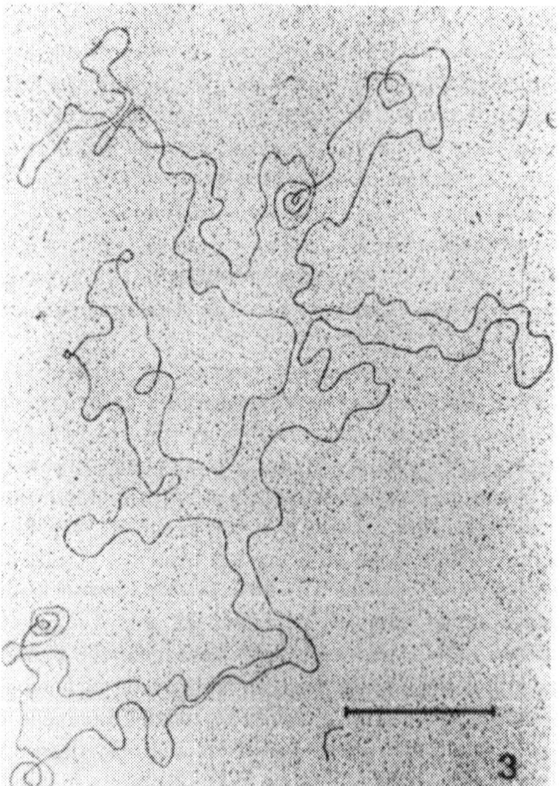

Figure 1. Electron micrograph of a circular DNA molecule from tobacco plastids. The bar indicates 1 μm. By kind permission from Seyer, P., Kowallik, K.V., and Herrmann, R.G. (1981) Current Genetics 3, 189–204, Elsevier Science.

in the nucleus and the lack of this methylated base in chloroplast DNA could also be used to distinguish chloroplast DNA from nuclear DNA.

Since DNA is sheared relatively easily during extraction, early size estimates of organelle DNA were prone to error. Kleinschmidt (1968) described a method of using a monomolecular layer of protein, floating on the surface of water, to adsorb DNA. Large DNA molecules are held intact in this two-dimensional protein mesh. They can be picked up on a grid, dried, surface-shadowed with platinum, and visualized under the electron microscope. Freshly prepared chicken liver mitochondrial DNA coated with cytochrome *c* was found to be composed predominantly of supercoiled circular molecules. The minor fraction of open circles had a mean contour length of 5.5 μm or 15 kb (van Bruggen et al., 1966). DNA extracted from tobacco chloroplasts also contains circular DNA molecules (Figure 1). Relaxed circles have a mean contour length of around 49 μm which corresponds to a circle of 160 kb (Seyer et al., 1981). Most mitochondrial and chloroplast genomes have a circular conformation. Exceptions include the mitochondrial genome of *C. reinhardtii* which is a 16 kb linear DNA molecule (Michaelis et al., 1990). Bacteria also contain circular genomes although they are much larger than those found in organelles. Sedimentation equilibrium measurements provided an alternative method for estimating the sizes of organelle genomes and were in general agreement with electron microscopic measurements.

The circular conformation of many organelle genomes provided an alternative route for purifying organelle DNA. Intercalating drugs such as propidium diiodide and ethidium bromide reduce the buoyant density of DNA. Since closed circular DNA molecules bind less drug than linear DNA molecules, organelle DNA bands below nuclear DNA on isopycnic caesium chloride gradients. Isolation of this lower band allowed the preparation of highly purified mitochondrial DNA (Smith et al., 1971) and chloroplast DNA (Kolodner and Tewari, 1975).

Coding Potential of Organelle Genomes

DNA reassociation experiments pioneered by Britten and Kohne (1968) for the analysis of genomes provided a route for determining the amount of unique DNA present in organelles (see also Chapter 2). The estimated amount of unique DNA calculated by this method is known as the kinetic complexity. Essentially the method operates on the observation that when the same amount of DNA is taken from two species, and then denatured, the subsequent renaturation rate is a function of genome size. Smaller genomes reassociate more quickly than larger genomes. Despite the variability inherent in this method, the kinetic complexities of organelle DNAs were in general agreement with the sizes estimated by electron microscopy and sedimentation studies (reviewed in Tewari, 1979). This suggested that the total coding potential of organelle DNA was present in circular DNA molecules composed of a single type of sequence. Organelle DNA can represent a significant percentage of total DNA. The small sizes of these genomes meant that they must be present in multiple copies per cell to account for this large contribution to total DNA. In *C. reinhardtii*, chloroplast DNA represents around 15% (80 copies/cell), while mitochondrial DNA represents less than 1% (50 copies/cell) of the total mass of DNA present in cells.

Chloroplasts and mitochondria contain their own transcription/translation machinery to decode their genomes. Their RNA polymerases and ribosomes are distinct from those present in the nucleus and cytosol. Chloroplast ribosomes have a sedimentaton coefficient of 70S compared with a value of 80S for cytosolic ribosomes (Lyttleton, 1962). Inhibitors of translation such as cycloheximide reduce cytosolic protein synthesis, but have little affect on mitochondrial or chloroplast protein synthesis. In contrast, chloramphenicol, an inhibitor of bacterial protein synthesis, prevents the synthesis of chloroplast (Ellis, 1977) and mitochondrial (Clark-Walker and Linnane, 1967) polypeptides, but has little affect on cytosolic protein synthesis. By comparing differences between normal, cycloheximide-treated and chloramphenicol-treated cells it was possible to assign the synthesis of proteins to cytosolic or organelle ribosomes. From these studies it became clear that the vast majority of mitochondrial and chloroplast proteins are encoded by genes located in the nucleus. Mitochondrial protein synthesis produces less than 10% of the proteins present in mitochondria. While all organelle genomes encode the RNA components of organelle ribosomes, most of the genes for ribosomal proteins are located in the nucleus.

Protein synthesis by isolated organelles provided an alternative approach for estimating the number of genes encoding polypeptides in organelle DNA. Isolated pea chloroplasts incubated with [^{35}S]-methionine use light as an energy source to synthesize approximately 80 polypeptides (Ellis, 1977). These can be visualized as radioactive spots on two-dimensional O'Farrell (1975) gels. Similar experiments have been carried out with isolated mitochondria using ATP as an energy source (Schatz and Mason, 1974; Grivell, 1983; Forde et al., 1978). Abundant organelle-encoded polypeptides such as the large subunit of ribulose bisphosphate carboxylase-oxygenase (Rubisco) of chloroplasts were among the first to be characterized. In tobacco, the gene encoding the large subunit of Rubisco is only inherited from the female parent. This is consistent with the maternal inheritance of plastid-specified traits that had been documented by Correns over 50 years earlier.

The limited coding potential of organelle DNA raised two important questions. First, how were polypeptides encoded by the nucleus and synthesized in the cytosol imported into organelles? Second, since functional organelles require the expression of genes located in different subcellular compartments, what coordinated the expression of genes that were physically separated in the cell? Multisubunit complexes composed of some organelle-encoded subunits with the remainder being encoded by the nucleus were of particular interest. Rubisco, which is composed of a plastid-encoded large subunit and nucleus-encoded small subunit, became a paradigm to explain this regulation (discussed later).

Protein Targeting and Complex Assembly

Unlike animal secretory proteins which are transported during their synthesis on membrane-bound ribosomes (see Chapter 5), the small subunit of Rubisco is completely synthesized in the cytoplasm and then transported into chloroplasts. This observation provided one of the first exceptions to the view, prevalent at the time, that cotranslation was necessary for protein transport across membranes (Ellis, 1979). The import of the small subunit of Rubisco into isolated chloroplasts requires an N-terminal transit peptide which is removed after entry into the chloroplast. Many nuclear-encoded mitochondrial proteins are also synthesized as precursors with an N-terminal presequence that is necessary for import into mitochondria (Neupert and Schatz, 1981).

Studies on the assembly of Rubisco holoenzyme, which is composed of eight large subunits and eight small subunits, showed the requirement for a large subunit binding protein. The binding protein is oligomeric and consists of two types of subunit $\alpha_6\beta_6$. It acts to prevent aggregation of large subunits and ensures their correct folding and assembly into Rubisco holoenzyme. It acts as a "molecular chaperone" (Ellis, 1987), a role first ascribed to nucleoplasmin (Laskey et al., 1978). This popular term has since grown to encompass an expanding list of proteins (Ellis and van der Vies, 1991). Chaperonins are homologous chaperone proteins present in bacteria, mitochondria, and plastids. The large subunit binding protein for Rubisco was the first chaperonin to be described.

DETAILED CHARACTERIZATION OF ORGANELLE DNA

The availability of restriction enzymes and recombinant DNA techniques in the 1970s and early 1980s revolutionized the study of extranuclear DNA. Restriction enzyme digests of purified mitochondrial and chloroplast DNA allowed precise characterization of their sizes and detailed cleavage maps to be produced. Circular restriction enzyme maps were produced in many instances including the 16.6 kb mitochondrial genome of humans (Brown and Vinograd, 1974) and 132 kb chloroplast genome of maize (Bedbrook and Bogorad, 1976). The mapping data also showed that organelle DNA was largely deficient in repeated DNA sequences in agreement with much of the kinetic complexity data. However, maize chloroplast DNA is representative of a large number of land plant chloroplast genomes in possessing two copies of a sequence (usually 20–30 kb) which are inversely oriented with respect to each other.

Cloning vectors of *Escherichia coli* allowed restriction fragments of chloroplast and mitochondrial DNA to be cloned and subjected to detailed analysis. These data located the sites of the cytoplasmic mutations studied by Ephrussi and Sager. Analysis of mitochondrial DNA from *petite* cytoplasmic mutants of yeast revealed gross rearrangements of mitochondrial DNA (Bernardi, 1979). A base change in the gene encoding the small ribosomal subunit protein S12 of plastid DNA is responsible for streptomycin resistance in Sager's *sr*-2 mutant (Liu et al., 1989). The non-Mendelian inheritance of white sectors provided early evidence for the presence of extranuclear DNA in plants. Mutations probably account for some but not all cases of green/white variegation in plants. Loss of functional plastid ribosomes through physical shock or deleterious nuclear loci will also produce permanently bleached plastids. If plastid ribosomes are lost, plastid-encoded ribosomal proteins cannot be translated to reconstitute functional ribosomes (Walbot and Coe, 1979). Ribosome-free plastids containing an intact genome can still divide and are inherited in a non-Mendelian fashion.

The advent of rapid DNA sequencing methods developed by Sanger[12] and colleagues in Cambridge provided a key to unlock the coding content of organelle DNA. The complete sequences of the 16,659 bp human and 16,295 bp mouse mitochondrial DNAs were published in 1981 (Anderson et al., 1981; Bibb et al., 1981). An important early outcome from sequence analysis of the mammalian and yeast mitochondrial genomes was the observation that the genetic code was not universal (Borst and Grivell, 1981). The stop codon UGA in the standard genetic code is read as tryptophan in mammalian and yeast mitochondria. The codons AGA and AGG represent arginine in the standard code but represent a stop codon in mammalian mitochondria. Plant mitochondria appear to use the standard genetic code (Walbot, 1991). Plastid genomes are much larger than animal mitochondrial genomes. The first plastid genomes, the 155,844 bp plastid genome from tobacco and 121,024 bp liverwort plastid genome, were not sequenced until 1986 (reviewed in Shimada and Sugiura, 1991).

ORIGINS OF MITOCHONDRIA AND PLASTIDS

Today over 20 mitochondrial genomes, from fungi, nematodes, insects, sea urchins, vertebrates, green algae, and land plants, and 11 plastid genomes have been sequenced and deposited in the European Molecular Biology Laboratory Nucleic Acid data base. In addition, partial organelle DNA sequences from a large number of additional species are available for analysis. These DNA sequences provide a wealth of data for comparative analyses between mitochondrial, plastid, and nuclear genes. These comparisons have provided a plausible answer to a long-standing question on the origin of mitochondria and plastids. Did mitochondria and plastids originate from within, by internal compartmentation, or from outside the cell? An origin from outside the cell was implied from the endosymbiosis model. In this model present-day mitochondria and plastids are the descendants of free-living bacteria that entered an ancestral proto-eukaryotic cell. The idea that organelles were derived from bacteria is old (Schimper, 1885; Mereschkowsky, 1905) and was based on similarities between bacteria and organelles observed under the microscope (reviewed in Taylor, 1970). The similarities between the structure and mode of expression of prokaryotic and organellar genes strongly supports an endosymbiotic origin for plastids and mitochondria. The resemblance between *E.*

coli and plastid genes is particularly striking. The 16S ribosomal RNA genes of maize plastids and *E. coli* share 74% identity (Schwarz and Kössel, 1980). Moreover, since plastid expression signals function in *E. coli*, plastid genes can be expressed in *E. coli* without modification (Gatenby et al., 1981).

Detailed phylogenetic comparisons between DNA sequences, mainly ribosomal genes, present in organelles and those present in bacterial DNA have provided clues on organelle ancestry. Plastids are likely to be the descendants of cyanobacteria, while animal mitochondria appear to be derived from the α-subdivision of the purple sulphur bacteria (Gray, 1993). Different mitochondrial genomes encode a similar set of functional proteins, but exhibit a high degree of diversity in their sizes, structural organization, and modes of expression. The conserved set of proteins in mitochondria suggests diversification from a common origin. New data on sequencing mitochondrial DNA of a red alga suggest plant and animal mitochondria share common ancestry (Leblanc et al., 1995). The dramatic differences between mitochondrial genomes will be illustrated by taking examples from mammals, fungi, and plants. While it is clear that plastids in green algae, such as *Chlamydomonas*, and land plants are derived from a single endosymbiotic event, the question of whether the plastids of *Euglena* and red algae are derived from separate endosymbiotic events is still unresolved (Palmer, 1993). Here we will restrict our detailed description to the plastids present in green algae and land plants.

MITOCHONDRIAL DNA

Mammalian Mitochondrial DNA: A Compact Genome with an Unusual Expression Pattern

The human mitochondrial genome was sequenced in 1981 by Sanger's research team while piloting his new dideoxy sequencing method (Anderson et al., 1981)(see Chapter 3). In the current era of direct sequencing of polymerase chain reaction products, it is salutory to consider that graduate students on his team may have sequenced only a few kb each as the major part of a doctoral thesis. The genome is extremely compact so that almost every base is part of the coding sequence. Each copy of mitochondrial DNA encodes 13 protein coding genes, 22 transfer RNAs (tRNAs), and 2 ribosomal RNAs (Figure 2A). Cytosolic protein synthesis requires a minimum of 31 tRNAs. In order for 22 tRNAs to cover the 61 possible codons the normal codon-anticodon pairing rules are relaxed in mitochondria so that many tRNAs accept any one of the four bases in the third wobble position. This allows one tRNA to pair with four possible codons. In these tRNAs, U in the first position of the anticodon allows U:N wobble. The discovery that mammalian mitochondria use their own genetic code altered our thinking of the universality of the standard genetic code common to bacteria, nuclei, and plastids. These changes in the code are thought to have occurred by random genetic drift. A small genome like mammalian mitochondrial DNA encodes only a small number of proteins. A change

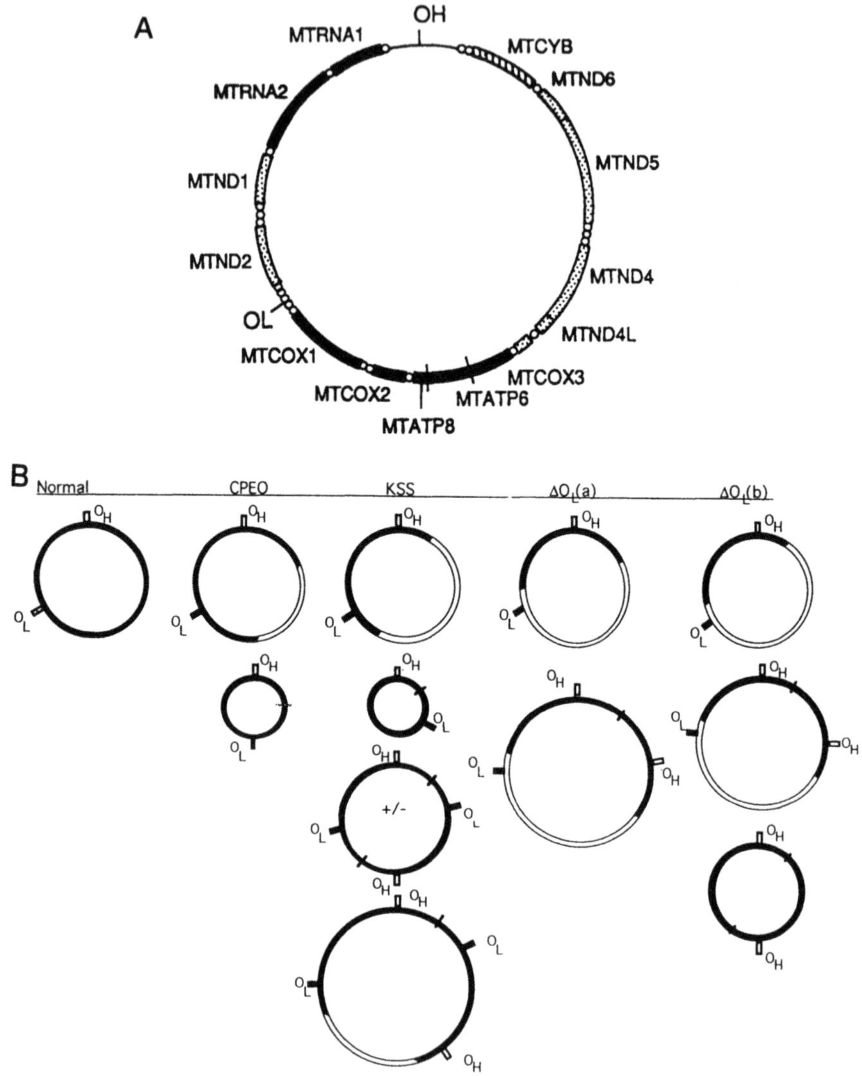

Figure 2. (**A**) The human mitochondrial genome.Origin of replication of the heavy strand (O_H). The light strand origin (O_L) is deleted in two cases, Δ O_L (**a**) (Dunbar et al., 1993) (**b**) (Ballinger et al., 1994; Ballinger et al., 1992). Circles represent genes for transfer RNAs, MTATP6, and MTATP8: subunits 6 and 8 of ATP synthase, MTCOXI to III; subunits I to III of cytochrome oxidase, MTCYB; cytochrome *b*, MTND1 to 6 and 4L; subunits 1 to 6 and 4L of NADH dehydrogenase, MTRNA1, and 2; 12S and 16S ribosomal RNAs. Shading corresponds to Figure 1b. (**B**) Rearranged molecules found in mitochondrial disease. Each column illustrates the different but related mtDNA rearrangements found in CPEO, KSS, and in families with deleted light strand origins, ΔO_L(a) and (b). Wild-type mtDNA is present in all cases. Deletion dimers are present in some but not all patients with KSS (this is indicated by "+/–"). Positions of deleted segments are white for each rearrangement.

in the specificity of a tRNA could result in a few amino acid substitutions with minor effects on protein function. In a large genome, encoding many proteins, deleterious amino acid substitutions would occur in some proteins and would be lethal to the cell.

Many of the proteins encoded by mitochondrial DNA were identified by raising antibodies against synthetic peptides predicted by the DNA sequence. The genome encodes subunits of the electron transport chain—namely, subunits 1 to 6 and 4L of complex I (Chomyn et al., 1985), apocytochrome *b* from complex III, subunits I to III of complex IV, and subunits 6 and 8 of ATP synthase as shown in Figure 2A. The mitochondrial genome is extremely compact with most of the genes abutting, and indeed ATPase 6 overlaps ATPase 8 and ND4 with ND4L. There is only just over 1 kb of noncoding DNA, most of it in the D-loop region (Figure 2A). This is an important control region, containing both the origins of replication of mitochondrial DNA (O_H and O_L from which synthesis of the daughter light and heavy strands start), and the promoters for RNA synthesis (LSP and HSP corresponding to light and heavy strand transcripts). The internationally accepted abbreviations for mitochondrial genes are listed in the legend of Figure 2A.

Transcription of Human Mitochondrial DNA

Human mitochondrial DNA genes are transcribed from LSP and HSP into two long polycistronic RNAs that diverge from each other. The two ribosomal RNA genes, 14 tRNAs, and 12 protein coding genes are in the primary transcript of the H-strand, while 1 protein and 8 tRNA genes are encoded by the transcript of the L-strand (see Attardi, 1985). The nuclear-encoded mitochondrial RNA polymerase requires a specificity factor, mitochondrial TFA, which has been purified, characterized, and cloned (Parisi et al., 1993). Mammalian mRNA species are unusual in lacking 5′ and 3′ untranslated regions. Moreover, over half lack a complete stop codon. A partial stop codon comprised of U or UA at the 3′ terminus is converted into UAA by polyadenylation.

Attardi noted that the tRNA genes are not randomly distributed about the genome. Rather, tRNA genes are found to intervene between ribosomal RNA and protein coding genes. The tRNA genes lie either immediately adjacent to one side, or more frequently to both sides, of ribosomal RNA and protein coding genes. Attardi suggested that the tRNAs act as punctuation marks in the primary H-strand transcript that allow processing into the mature RNAs. In the 15 years following the punctuation model, mitochondrial RNA transcription and processing remains a poorly understood area.

Replication of Human Mitochondrial DNA

There are two origins of replication: the origin for the heavy strand (O_H) in the D-loop, and the origin for the light strand (O_L) in a small noncoding region in a cluster of tRNAs approximately 8 kb away. Replication of mitochondrial DNA is initiated at O_H, by a γ DNA polymerase, primed by an RNA transcript initiated at the LSP. Those transcripts which are not destined to become mRNA are cleaved by the mitochondrial RNA processing activity (RNAse MRP) exactly at the position corresponding to the 5′ end of the nascent DNA strand. Interestingly, the RNA component of this endoribonuclease (MRP) is highly homologous with the nuclear RNAse P. As the vast majority of MRP is actually extra-mitochondrial it may have an additional role such as processing nuclear ribosomal RNAs. Similarly, mitochondrial TFA has features in common with the ribosomal transcription factor HUBF (Jantzen et al., 1990) and could have an additional role. It is possible that these two nuclear genes for mitochondrial biogenesis evolved from the corresponding control elements for nuclear ribosomal RNA.

Replication of the light strand is not initiated until the replication fork has gone around 2/3 of the genome and reaches O_L, leaving the light strand single stranded until this occurs. It is interesting that the bias in the light strand towards high C content and low G content increases with the distance from O_L, and it may be that the guanines in the light strand are particularly susceptible to mutation while it remains single stranded. A similar distribution of this bias in relation to origin of replications has been observed in many species. Once the replication fork has reached O_L, the origin is probably made available to the DNA polymerase by the formation of a hairpin-like structure. DNA synthesis is primed by a short stretch of riboadenosines synthesized by an RNA primase. The synthesis of the daughter strands has been estimated to take about one hour, during which time there is a variable length of heavy strand left unpaired and potentially vulnerable to recombination. Gapped daughter molecules are then converted to closed circles, and negative superhelical turns introduced by a DNA gyrase.

Yeast Mitochondrial Introns Encode Proteins that Allow Self-Splicing and Mobility

The mitochondrial DNA of bakers' yeast, *Saccharomyces cerevisiae*, is 85 kb in size yet it only encodes a similar number of products to mammalian mitochondrial DNA: three subunits of cytochrome *c* oxidase, cytochrome *b*, three ATPase subunits, one ribosomal protein, 22 tRNAs, as well as ribosomal RNAs (deZamaroczy and Bernardi, 1986). The increase in size of yeast mitochondrial DNA relative to mammalian DNA is due in part to AT-rich intergenic spacer DNA. Apart from the 15S and 21S ribosomal RNA genes, which are processed from a single precursor transcript, the protein coding genes are transcribed from separate promoters. In contrast to mammalian mitochondrial mRNAs, yeast mitochondrial mRNAs have 5′ and 3′ untranslated regions and are not polyadenylated. This mitochondrial genome has not been subjected to the same extreme evolutionary pressures that resulted in a small mitochondrial genome in mammals.

One of the surprising features was the presence of intervening sequences or introns within mitochondrial genes. Introns had just been discovered in nuclear genes and were absent in bacterial genes. Given the prokaryotic nature of the mitochondrial expression system, the presence of intron sequences, a characteristic of eukaryotes, was unexpected. A further novel observation of some yeast introns was that they contained open reading frames necessary for the splicing reaction (Lazowska et al., 1980). Mutations in the intron-located open reading frames abolished splicing. The cytochrome *b* gene in some strains of yeast is interrupted by five introns while other strains lack the first three introns. Both long and short genes produce the correctly spliced mRNA coding for cytochrome *b*. Removal of the first intron fuses exons one and two to the second intron in an intermediate RNA. This creates an open reading frame composed of 143 amino acids from exons one and two followed by 250 amino acids from intron two at the C-terminus. The resulting chimeric protein, called a maturase, is required for splicing out intron two. By facilitating the removal of intron two the maturase destroys its own synthesis. This manner of regulating mRNA processing is novel and appears to be unique to fungal mitochondria.

The introns present in the yeast mitochondrial genome are distinct from those present in most eukaryotic pre-mRNAs. The 5′ and 3′ termini of these introns do not conform to the GU..AG rule of most nuclear introns. Michel and Dujon (1983) recognized two classes of organelle introns based on their extensive secondary structure, both of which have been found in mitochondria and plastids. Group I introns are also found in the nucleus of *Tetrahymena thermophila* and the T4 bacteriophage of *E. coli*. A number of group I and group II introns can self-splice *in vitro*. (Tabak and Grivell, 1986). Self-splicing via RNA catalysis may represent an ancient process that preceded protein catalysis. Many Group I introns contain maturases (see above) necessary for *in vivo* splicing or an internal open reading frame encoding an endonuclease (see below). Some Group II introns encode a reverse transcriptase. In yeast, nuclear-encoded aminoacyl-tRNA synthetases have a dual function. Besides charging tRNAs with amino acids they are also involved in splicing mitochondrial genes (Benne, 1988).

Studies on a Group I intron in the 21S ribosomal gene of yeast mitochondria showed that it was a mobile. Crosses between yeast strains containing the intron ($omega^+$) strains with those lacking the intron ($omega^-$) result in progeny that almost exclusively contain an intron in the 21S RNA gene. In this case the intron encodes an endonuclease that cleaves mitochondrial DNA at its 18 bp recognition site which lies in the uninterrupted 21S ribosomal gene. Repair of this double-stranded DNA break uses the intron-containing 21S ribosomal gene as a template to bridge the gap (double strand break repair model). This results in the insertion of the intron into the previously uninterrupted gene. The recognition site of the endonuclease is destroyed by insertion of the intron. This makes intron insertion irreversible and the intron is ready to insert itself into more intronless versions of the 21S ribosomal RNA gene. This process is called intron homing (Dujon et al., 1989).

Plasmids sharing little or no homology with the main mitochondrial genome have been found in the mitochondria of filamentous fungi. The 3581 bp Mauriceville plasmid from *Neurospora crassa* has been completely sequenced. It encodes a protein related to reverse transcriptase and appears to be replicated via a full-length RNA transcript that is reverse transcribed into DNA. As such the plasmid exhibits the same replication strategy as animal retroviruses. The presence of reverse transcriptase in Group II introns raises the possibility that the Mauriceville plasmid is related to the progenitors of present DNA mitochondrial introns (Narang et al., 1984).

RNA Editing in Mitochondria

Colinearity between the amino acids of polypeptides and the genetic code was deduced from elegant experiments in the early 1960s. Alteration of nucleotides in mRNA so that it no longer corresponded to its DNA template came as a complete surprise. The insertion of nucleotides into RNA is called editing and it creates initiation codons, extends open reading frames, and creates termination codons (Weiner and Maizels, 1990). The process of RNA editing was discovered in the mitochondria of *Trypanosoma brucei* and *Crithidia fasiculata* where it was shown that transcripts of cytochrome *c* oxidase subunit II (COII) gene contain four uridines near their 5′ ends that are not encoded by the gene. The presence of these extra uridines suppress internal frameshifts in the genomic sequence, thereby creating a complete open reading frame. In the case of the cytochrome *c* oxidase subunit III (COIII) gene, almost 60% of the nucleotides in the sequenced mRNA are edited nucleotides that are not present in the gene. RNA editing challenged our basic understanding of the rules governing the storage and expression of genetic information. If the DNA sequence of a template can be changed by RNA editing, what governs the specificity of editing?

Kinetoplastid protozoans derive their name from the organization of their mitochondrial genome. Their solitary mitochondrion contains 20–50 copies of homoplasmic maxicircles and 5000–10,000 copies of heterogeneous minicircles which are concatenated into a single dense network called the kinetoplast. Small RNA molecules (guide RNAs) encoded by both maxicircles and minicircles are complementary to the mRNAs encoding mitochondrial polypeptides. These hypothetical duplexes contain G:U pairing. These guide RNA sequences could provide the specificity required for editing (Weiner and Maizels, 1990).

Plant mitochondrial transcripts are also complex. Each gene produces multiple transcripts which are initiated 100 bases to several kb upstream of the coding region. All protein coding transcripts in higher plant mitochondria that have been examined are edited (Walbot, 1991). In some transcripts such as the nad3 transcript of wheat, 13.5% of the amino acid residues predicted from the DNA sequence are altered in the mRNA. The majority of editing events involve C to U changes but U to C changes have also been detected. A cytoplasmic male sterile line in wheat has defects in editing *atp*9 mRNA. A dominant restorer gene located in the nucleus, that confers fertility, restores normal editing of *atp*9 mRNA. Isolation of the restorer gene and elucidation of its product may provide some information on the mechanism of editing in plants. Editing has also been observed in plastids.

PLANT MITOCHONDRIAL GENOMES

Plant Mitochondrial Genomes are Large, with Multipartite Genomes and Small Plasmids

The largest mitochondrial genomes are found in higher plants ranging from 218 kb in turnip to around 2500 kb in melons (Levings and Brown, 1989). This increase in size is not associated with a dramatic increase in the number of mitochondrial genes. About 30 to 40 proteins are thought to be encoded by plant mitochondrial genomes. The mitochondrial genome of the liverwort *Marchantia polymorpha* represents a primitive form of plant mitochondrial genome and is comprised of a single DNA species. The 186,608-bp genome contains 27 tRNA genes, ribosomal RNA genes and 30 open reading frames (ORFs) coding for 16 ribosomal proteins, 3 subunits of ATPase, 3 subunits of cytochrome *c* oxidase, apocytochrome *b*, 7 subunits of NADH ubiquinone oxidoreductase, and 3 proteins of unknown function (Oda et al., 1992). Higher plant mitochondrial genomes are unusual in that they are often composed of more than one circular DNA species. The significance of this is unclear and some higher plants only possess a single circular DNA species.

Small circular or linear DNA plasmids have not only been found in filamentous fungi but also in plant mitochondria. They were first studied because of their possible involvement in cytoplasmic male sterility. Cytoplasmic male sterility prevents the production of pollen and is maternally inherited. It prevents self-fertilization of flowers and as such is useful to plant breeders who would otherwise have to employ workers to remove anthers by hand. Recent work has shown that the main mitochondrial genome and not plasmids are responsible for male sterility (see below). Two linear dsDNA plasmids, S1 (6.4 kb) and S2 (5.4 kb), are found in the mitochondria of cytoplasmic male sterile maize plants containing S cytoplasm (Levings and Brown, 1989). These linear plasmids resemble the adenovirus genome in containing terminal inverted repeats capped by terminal proteins. In adenovirus the terminal proteins are implicated in the initiation of DNA replication. Sequences resembling the terminal repeat of S1 and S2 are present in the main mitochondrial genome. Recombination between the linear plasmids and the main genome at these sites linearizes and fragments the main genome. Fragmentation into linear subgenomic DNA species does not result in loss of any essential genes and is not permanent. Loss of S1 and S2 results in restoration of a circular genome. The plasticity of plant mitochondrial genomes is bewildering. Single- and double-stranded RNAs have also been found in the mitochondria of maize. These RNA plasmids lack homology with mitochondrial DNA and their origin is not clear. It is conceivable that they are descendants of an RNA virus (Lonsdale, 1986).

Scrambled Ribosomal RNA Genes

Most large and small subunit cytoplasmic ribosomal RNAs are synthesized as continuous polyribonucleotide chains (see Chapter 5). In the mitochondrial genomes of *C. reinhardtii* and *Tetrahymena pyriformis* the sequences of the ribosomal RNA genes are not colinear with genes producing continuous ribosomal RNAs. For example, in *C. reinhardtii* mitochondria, the small and large subunit ribosomal RNA coding sequences are interspersed among each other (Boer and Gray, 1988). The different interspersed RNA coding sequences are expressed and processed into separate low molecular weight RNA species. The extensive secondary structure of ribosomal RNA together with protein–RNA interactions allows the correct assembly of fragmented RNAs into a cohesive unit.

Promiscuous DNA

Plant mitochondrial genomes contain insertions of plastid DNA. The first characterized sequence was a 12 kb insertion of plastid DNA into maize mitochondrial DNA (Stern and Lonsdale, 1982). The sequence contained the genes for plastid 16S ribosomal RNA and tRNAs. Until this discovery, the DNA in plastids and mitochondria were thought to be evolving in complete isolation. DNA transfer between cellular compartments gave rise to the term "promiscuous DNA" (Ellis, 1982). Analysis of rice mitochondrial DNA reveals 22 kb of plastid DNA scattered around the 492-kb genome (Nakazono and Hirai, 1993). Higher plant mitochondrial genomes, which contain large amounts of plastid-derived DNA, contrast with the liverwort mitochondrial genome, which lacks any foreign DNA sequences.

Mitochondria Import RNA

The discovery that the RNA primer used to initiate DNA synthesis in mammalian mitochondria was imported from the nucleus dispelled the belief that only proteins but not RNA could cross organelle membranes (Chang and Clayton, 1987). Evidence that tRNAs are imported from the cytosol into mitochondria is derived from elucidating the coding capacity of mitochondrial DNA and also characterizing tRNAs present in isolated mitochondria. Mammalian mitochondria use 22 unusual tRNAs to decode the 61 sense codons. The linear 15.8-kb genome of *C. reinhardtii* only encodes three tRNAs (Michaelis et al., 1990). Although liverwort mitochondrial DNA encodes 27 tRNA species, two species necessary to read leucine and threonine codons are absent. *C. reinhardtii*, plant, and trypanosome mitochondria appear to import nuclear-encoded tRNAs to make up a complete set for protein synthesis. Eleven of the 31 tRNA species present in potato mitochondria are encoded by nuclear DNA and are imported from the cytosol (Dietrich et al., 1992).

The plastid genome of *Epifagus virginiana*, a nonphotosynthetic parasite of beech trees, lacks 13 tRNA genes found in green plastids. This suggests tRNA import can also occur in plastids (Wolfe et al., 1992).

ORGANIZATION AND EXPRESSION OF PLASTID DNA

In contrast to mitochondrial genomes, the plastid genome is relatively well conserved in land plants and green algae. It is a compact genome, packed with genes varying from 120–200 kbp. Many plastid genomes contain a large inverted repeat (usually 20–30 kb) that encode the 23S and 16S ribosomal RNAs. In land plants, this large inverted repeat divides the genome asymmetrically into large and small single-copy regions. Recombination across the inverted repeat inverts the relative orientation of single-copy regions. Restriction mapping revealed equal numbers of both inversion isomers. This suggests frequent inversion of the single-copy regions by recombination events between the inverted repeats during plant growth (Palmer, 1985). The presence of the large inverted repeat is associated with conservation of gene arrangement such that the same gene order is often found in different vascular plants. Changes in gene order mainly result from inversions of DNA in the large single-copy region. The order of genes is highly rearranged in plastid genomes such as pea that lack a large inverted repeat organization (Palmer, 1985). Homoplasmy (all genomes identical in an individual) is the norm and the plastids of land plants lack plasmids.

The gene content and organization of the rice, tobacco, and liverwort plastid genomes are very similar (Palmer, 1990; Shimada and Sugiura, 1991). Gene order in rice is changed relative to tobacco by three overlapping inversions in the large single-copy region. The 134,525 bp rice plastid genome was the third plastid genome to be sequenced in entirety after liverwort and tobacco (Hiratsuka et al., 1989, Figure 3). It illustrates the organization and gene content of land plant plastid genomes. The 20.8 kb large inverted repeat (IR) separates a 80.6 kb large single-copy region (LSC) from a 12.3 kb small single-copy region (SSC). Genes in the large inverted repeat are present in two copies per genome. It encodes 30 different tRNAs (trn genes) which follow the standard genetic code, about 80 polypeptides, and the ribosomal RNAs (16S, 23S 4.5S, 5S). A tRNA charged with glutamate has a dual role in plastids. Apart from its role in protein synthesis glutamyl-tRNA also serves as a precursor for chlorophyll synthesis (Schön et al., 1986). Most of the genome codes for photosynthetic proteins or components of the transcription/translation system. Most of the ribosomal proteins and RNA polymerase subunits were identified by comparison with similar proteins present in *E. coli*. The presence of eight ORFs (ndhA to ndh G, psbG) that resembled the respiratory-chain NADH dehydrogenase from human mitochondria was unexpected. Do these proteins have a role in plastids or are they exported to mitochondria? The absence of these genes from the plastid genome of a non-photosynthetic plant, the parasite *Epifagus*, suggests that they are more likely to be involved in photosynthetic metabolism

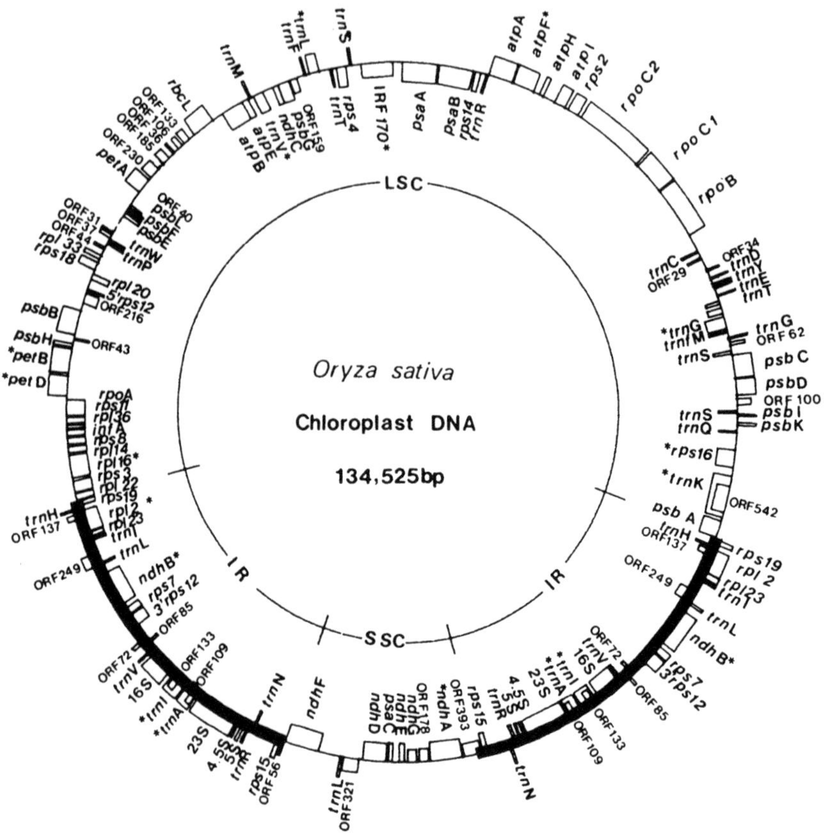

Figure 3. The gene organization of rice (*Oryza sativa*) chloroplast DNA deduced from its DNA sequence. Genes on the outside of the circle are transcribed counterclockwise and genes located inside the circle are transcribed clockwise. The locations of introns are denoted by asterisks. LSC: large single copy region. IR: inverted repeat. SSC: small single copy region. For further information see text. With kind permission from Hiratsuka et al., Figure 1: The complete sequence of the rice chloroplast genome (1989) Mol. Gen. Genet. 217, 185–194. Copyright notice 30 June 1995 of Springer-Verlag.

within the plastid (dePamphillis and Palmer, 1990). The function of a number of ORFs are still unknown.

The plastid genes of land plants have complex transcription patterns. Posttranscriptional and posttranslational events play a major role in gene expression. Some plastid genes such as that encoding the large subunit of Rubisco are transcribed into monocistronic mRNAs, while others such as atp F and atp A are found in polycistonic transcripts. *trans*-Splicing, a variation of the common *cis*-splicing, was first discovered in the nuclear genes of trypanosomes, and has also been observed in plastids. Plant mitochondria are known to contain a large number of *trans*-spliced mRNAs (Wissinger et al., 1992). In *C. reinhardtii* plastids a small RNA molecule pairs with the introns of two *trans*-spliced precursor RNAs to form a structure that resembles a group II intron (Goldschmidt-Clermont, 1991). A group II intron appears to have been fragmented into three RNA species. This discovery has implications for our general understanding of splicing. The small nuclear RNAs in the spliceosome could be derived from an ancient self-splicing intron (Sharp, 1991). RNA editing, a feature of plant mitochondria, also take places in plastids (Hoch et al., 1991).

The genes present in plastids are largely conserved in land plants and green algae. Free living bacteria such as *E. coli* contain genomes of around 5000 kbp, much larger than the 120–200 kbp plastid genomes of green algae and land plants. The similarity in gene content of plastid genomes suggests that the primary endosymbiont suffered a dramatic reduction in genome size early in evolution. Genes required for a free-living existence were lost and a large number of genes were transferred to the nucleus. The *tuf*A gene encodes protein synthesis elongation factor TU and provides an example of a gene that has migrated to the nucleus, since the divergence of green algae and land plants, about 450–500 million years ago (Baldauf and Palmer, 1990). *tuf*A is found in the plastid genome of green algae but is absent in land plant plastid genomes. The question, "Why haven't all the genes in plastids been relocated to the nucleus?", remains unanswered.

RELOCATION OF ORGANELLE GENES TO THE NUCLEUS

The majority of polypeptides found in mitochondria and plastids are encoded by genes located in the nucleus. Over 95% of mitochondrial and 80–90% of plastid polypeptides are encoded by the nucleus (Palmer, 1990). The bacterial origins of plastids and mitochondria suggests that most of these nuclear genes were once present in the genomes of the endosymbiont. Evidence to suggest that nuclear genes for organelle-targeted polypeptides were once encoded by the genome of endosymbionts comes from comparing isozymes that are targeted to different cellular compartments. A comparison between the genes encoding cytosolic and plastidic forms of tobacco glyceraldehyde-3-phosphate dehydrogenase (GAPDH) shows that the genes are not closely related and could not have arisen as a consequence of duplication of an ancestral nuclear gene (Shih et al., 1986). The tobacco cytosolic form is more closely related to the animal GAPDH than it is to the plastidic form. The plastidic form evolved from a different lineage, presumably in the endosymbiont, and was transferred to the nucleus. Similarly, when nuclear genes encoding isozymes of vertebrate superoxide dismutase are compared, the mitochondrial enzyme resembles bacterial enzymes more than it resembles the cytosolic form.

Evidence for gene mobility between intracellular genomes comes from the observation that mitochondrial and plastid DNA can be found in other cellular compartments. Insertions of plastid DNA into plant mitochondrial DNA were first described in maize (Stern and Lonsdale, 1982). This was shortly followed by the observation that plastid DNA sequences were present in nuclear DNA (Timmis and Scott, 1983). Ongoing integration of DNA from plastids into nuclear DNA introduces the entire organelle genome, as small scattered DNA pieces, into the nuclear genome. Mitochondrial DNA has also been found in the nucleus (Farrelly and Butow, 1983). The presence of plastid and mitochondrial DNA in the nucleus has now been documented in a variety of species.

Laboratory experiments with bakers' yeast have allowed genes normally in the nucleus to be transferred into mitochondria. The *URA*3 marker gene can be introduced into the mitochondrial genome of yeast cells that contain a nonfunctional *ura*3 gene in the nucleus. Such strains require uracil in the medium to grow since the *URA*3 marker gene contains nuclear expression signals and does not function in mitochondria. Transfer of the functional *URA*3 marker gene from the mitochondrion to the nucleus allows the strain to grow in the absence of uracil. The number of URA+ prototrophs obtained provides an estimate of the rate of transfer of the *URA*3 marker gene to the nucleus. The rate of transfer is surprisingly high and was estimated to be approximately 2×10^{-5} per cell per generation (Thorsness and Fox, 1990).

In the yeast laboratory experiments the *URA*3 marker gene was already endowed with eukaryotic expression signals that would allow expression once it was transferred to the nucleus. Genes normally resident in the plastid or mitochondrion are flanked by regulatory regions that would function inefficiently in the nuclear/cytosolic compartment. In order for these genes to function in the nucleus they require eukaryotic expression signals. When organelle genes are first transferred to the nucleus they are unlikely to be functional. While a functional copy of the gene is retained in organelle DNA this is unlikely to be deleterious. Over evolutionary time, organelle genes relocated to the nucleus would acquire eukaryotic expression signals and an organelle targeting sequence to enable them to function. Once a functional copy of the organelle-derived gene is present in the nucleus, the original gene located in the organelle is no longer essential to the cell and can acquire mutations that result in its dysfunction and eventual loss. A situation resembling this hypothetical scheme has been described in fungi. The ATPase subunit 9 gene is encoded by the mitochondrial genome of bakers' yeast but by the nuclear genome of *Neurospora crassa*. However, a silent copy of the ATPase subunit 9 gene is present in the *Neurospora crassa* mitochondrial genome (van den Boogaart et al., 1982).

REGULATORY INTERACTIONS BETWEEN NUCLEUS AND ORGANELLE

The importance of nucleus–cytoplasm interactions has been known since the work of Renner (1922) on *Oenothera* plants (reviewed by Kutzelnigg and Stubbe, 1974).

Oenothera exhibits biparental inheritance of plastids. Hybrid bleaching, which leads to a green/white variegated phenotype, results from crosses between different *Oenothera* species if one of the two parental plastids is not compatible with the nucleus.

The numbers of plastids and mitochondria vary dramatically between different cell types in multicellular organisms. This is associated with large fluctuations in organelle DNA levels. In barley, plastid DNA levels vary from around 200 copies/cell in roots to over 10,000 copies/cell in young green leaves. In contrast, the amount of nuclear DNA remains relatively constant during growth and differentiation. Increases in plastid DNA levels are thought to be important to drive the rapid synthesis of plastid-encoded ribosomal RNAs which are required to assemble plastid ribosomes (see Chapter 2, Reiterated nuclear rRNA genes). Multi-enzyme complexes in organelles contain some subunits encoded by the nuclear genome while others subunits are synthesized from genes that are located in organelle DNA. Assembly of a functional complex requires the supply of subunits from the nuclear–cytosolic and organelle compartments. How is the expression of nuclear genes encoding organelle polypeptides coordinated with a vast excess of organelle genes? To answer this question, the synthesis of a nuclear or organelle-encoded product is inhibited and the effects of this disruption on the synthesis of proteins or mRNA encoded by the other organelle followed.

The effects of organelle mutations on nuclear gene expression has been studied in mitochondria and plastids. The mitochondrial ribosomes of bakers' yeast are assembled from ribosomal RNAs encoded by the mitochondrial genome and ribosomal proteins encoded by the nucleus. Nuclear encoded ribosomal proteins still accumulate in the mitochondria of *petite* mutations that are deficient in ribosomal RNAs. In this case, nuclear gene expression is not downregulated in response to impaired mitochondrial function (Shyjan and Butow, 1993). A nonsense mutation in the plastid gene for the large subunit of Rubisco prevents its synthesis in *C. reinhardtii*. The nuclear-encoded small subunit of Rubisco is still synthesized and imported into plastids. The absence of the large subunit prevents the assembly of imported small subunit into holoenzyme and it is degraded in the plastid (Spreitzer et al., 1985). The stoichiometry of plastid and nuclear-encoded subunits of complexes appears to be regulated at the posttranslational level. Excess subunits that are not stabilized by incorporation into complexes are rapidly degraded. These two examples show that the absence of an organelle-encoded subunit does not always lead to downregulation of the synthesis of a nucleus-encoded organelle protein.

A number of other studies do demonstrate that the expression of nuclear-encoded organelle proteins are sensitive to the functional state of organelles. In *petite* mitochondrial mutants of bakers' yeast a number of nuclear genes are upregulated compared with wild-type cells. These include the gene encoding the mitochondrial isoform of citrate synthase (Shyjan and Butow, 1993). Inhibition of mitochondrial function in human cells with chloramphenicol leads to a decrease in the accumulation of the mRNA for the mitochondrial ADP/ATP translocator that is encoded by the nucleus. When chloroplasts are bleached by herbicides the expression of a number of nuclear genes for plastid polypeptides are affected. Cytosolic mRNAs for the chlorophyll *a*/*b* binding protein are reduced (Taylor, 1989). These data from yeast and animal mitochondria, and plastids, indicate that the expression of nuclear genes is sensitive to the functional state of organelles. The nature of the signal that is perceived by nuclear genes is not known. It could involve the production of a signal molecule in organelles that moves to the nucleus.

A large number of genes in the nucleus are concerned with organelle maintenance. In yeast, a rough estimate suggests that as much as 25% of all nuclear genes in yeast could be required for the biogenesis and inheritance of functional mitochondria (Shyjan and Butow, 1993). Approximately a third of yeast nuclear mutations affecting mitochondria are in genes whose products regulate specific mitochondrial genes at the post-transcriptional level (Fox, 1986). In *C. reinhardtii* the expression of a single plastid gene requires 5–10 nuclear gene products. This could represent as much as 5–10% of the nuclear genome (Rochaix, 1992). It is clear that the nucleus invests in a large number of genes in order to regulate organelle functions.

VEGETATIVE SEGREGATION, RECOMBINATION, AND HOMOPLASMY

When plastids or plant mitochondria of different genetic constitution are mixed in the same cell they rapidly segregate into pure homoplasmic lines in the ensuing cell divisions. This rapid segregation of organelle genes into homoplasmic lines is known as vegetative segregation. It is responsible for green/white variegation in plants. Mixed populations of white and green plastids segregate away from each other during the cell divisions that accompany plant growth. Vegetative segregation of organelle genes is also observed in the descendants of somatic cell hybrids. Antibiotic-resistant mutations in plastid and mitochondrial DNA can be used to distinguish different populations of plastids and mitochondria. Both plastids and mitochondria exhibit vegetative segregation. When mitochondria of different genetic reconstitution are mixed in the same cytoplasm by cell fusion they exchange DNA sequences to produce recombinant mitochondrial genomes. This is true of plant (Belliard et al., 1979) mitochondrial genomes, but not of animal mitochondrial genomes (Yoneda et al., 1994; Hayashi et al., 1994b). Recombination between plastid genomes in *C. reinhardtii* has allowed the construction of a linkage map containing antibiotic resistance markers (Harris, 1989). In contrast recombination between different plastid genomes in higher plants is rare.

Vegetative segregation resolves mixed organelles into pure lines and results in homoplasmy of organelle DNA. In *C. reinhardtii* stable heteroplasmy of plastid genomes has been observed when two plastid DNA types are required for cell survival. Two instances have been described. In the first, a nonsense mutation in a plastid gene is suppressed by an altered tRNA (Yu and Spreitzer, 1992). The genes encoding the normal and altered tRNAs are alleles encoded by different plastid genomes. These genomes coexist in the same plastid since both tRNAs are required for cell viability. The second instance occurs when essential plastid genes are disrupted by antibiotic-resistant cassettes introduced into plastids by transformation. In the presence of the antibiotic both the disrupted gene expressing the antibiotic resistance and normal gene are required for cell viability (Goldschmidt-Clermont, 1991). Heteroplasmy is also associated with mitochondrial mutations in man (see below) and plastid mutations in cereal plants (Day and Ellis, 1984).

ORGANELLE DNA IS A USEFUL MOLECULAR CLOCK

Because recombination of mitochondrial DNA is a rarity (and indeed all the examples so far described in man have been associated with disease), the accumulation of mitochondrial DNA mutations has been used by population geneticists as a "molecular clock" (Cann et al., 1987). Each mitochondrial clone with its own set of mutations, inherited from its mother, evolves independently of other clones and can be used to deduce phylogenies. All mitochondrial DNA types so far described probably arise from a single African type around 200,000 years ago (the "mother Eve hypothesis," reviewed Poulton, 1987). This type of study shows that the majority of point mutations are "silent" because they occur in the third base position, that transitions occur more frequently than transversions, and that the mutation rate of coding regions of mitochondrial DNA is about 10 times that of nuclear genes. Mitochondrial DNA has a relatively high mutation rate probably because of free radical damage and rudimentary mitochondrial repair systems.

The rates of nucleotide substitution in plant organelle genomes are much lower than in animal mitochondria (Palmer, 1990). The rates of synonymous or silent base substitutions are approximately three times higher in plastid DNA than in plant mitochondrial DNA. In turn this rate of plastid mutation is on average one-fourth of that found in nuclear DNA from animals and plants. The conservation of gene arrangement and relatively low mutation rate has made plastid DNA a useful molecule for taxonomic studies in plants. Restriction enzyme polymorphisms resulting from base substitutions and small insertion/deletion events have been widely used to assess relatedness between species in different taxa. The analysis of DNA from fossils provides a particularly striking method to assess mutation rate and the validity of phylogenetic trees. Plastid DNA sequences have been obtained from a 17–20 million-year-old Magnolia leaf (Golenberg et al., 1990).

PHENOTYPES ASSOCIATED WITH ABNORMAL MITOCHONDRIAL DNA

The human mitochondrial genome was characterized and sequenced in 1981 (Anderson et al., 1981). Abnormalities in the genome are recognized as being responsible for human diseases with a maternal inheritance pattern and associated with defects of the respiratory chain, such as the mitochondrial encephalomyopathies. This is because mitochondrial DNA encodes peptides involved in electron transport (Attardi, 1985; Attardi and Schatz, 1988) and is maternally inherited (Giles et al., 1980). In general, these diseases affect tissues with a high energy demand, such as muscle, brain, and heart. Similarly, respiratory chain dysfunction may give rise to distinctive phenotypes in other organisms, such as the *petite* yeast strains discussed above. Pollen formation in higher plants has a high energy requirement, and consequently plants carrying mutant mitochondrial DNA strains may be male-sterile. Finally, respiratory chain function declines with age in both man and fungi in parallel with increasing levels of mutant mitochondrial DNA. Could the accumulation of defective mitochondria underlie senescence (see below)?

Mitochondrial DNA and Human Disease

Until 1988, the human mitochondrial genome had attracted little medical attention. While it was already a major player in population genetics, the African origin of mitochondrial Eve was an academic curiosity beyond the scope of core medical texts. Since then, discoveries of pathological mitochondrial DNA mutations have burgeoned, and have already reached double figures. A comprehensive review is therefore outside the scope of this chapter.

The mitochondrial encephalomyopathies are a group of diseases characterized by muscle weakness with an abnormal histological appearance. There is a massive proliferation of mitochondria which clump together in abnormal muscle cells. These can be stained to give a characteristic "ragged red" appearance. Although primarily affecting muscle, these disorders may have profound effects on brain, eyes, heart, endocrine organs, liver, kidney, pancreas, and blood. Both major rearrangements and point mutations of mitochondrial DNA have been identified, and in many cases the predictable respiratory chain dysfunction and maternal inheritance pattern is evident. However, several of the mitochondrial DNA diseases only fulfill one of these criteria: the first group of mitochondrial DNA diseases to be identified were sporadic deletions (Holt et al., 1988; Moraes et al., 1989), and the first point mutation, demonstrated in Leber's Hereditary Optic Neuropathy (LHON) (Wallace et al., 1988), is not associated with a clear biochemical defect. Furthermore, it is now clear that both autosomal dominant (Zeviani et al., 1989) and probably recessive (Moraes et al., 1991) nuclear genes can cause abnormalities of mitochondrial DNA. Examples of some of these conditions will be used to illustrate some of the major directions of current research.

Much of the variability of the mitochondrial myopathies may be attributable to heteroplasmy and/or interactions with isoforms of nuclear genes. Homoplasmy is the norm in healthy human beings. In disease, heteroplasmy (two or more distinct mitochondrial DNA populations within the same individual) is usual, presumably because homoplasmy for a severe mutation would be lethal to many types of cell. Indeed, it is such a common feature of the mitochondrial myopathies that some investigators have suggested that heteroplasmy should be one of the criteria used to determine whether or not a mitochondrial DNA mutation is pathogenic. The level of mutant mitochondrial DNA varies in different tissues, changes with time, and appears to be crucial in determining which tissue is involved and how the disease evolves. For instance, the presence of mitochondrial DNA duplications is associated with widespread disease manifesting as Kearn's-Sayre syndrome (KSS) (a syndrome of muscle weakness particularly affecting eye movements, retinal degeneration, cardiac conduction defects, and sometimes diabetes, deafness, and incoordination) with the mutations widely distributed in different tissues. More localized disease is associated with focal proliferation of mitochondrial DNA mutants as in chronic progressive external ophthalmoplegia (CPEO) (probably a mild variety of KSS). Whether and how mitochondrial DNA duplications affect mitochondrial DNA segregation is unknown.

In the early literature on mitochondrial diseases, some investigators asserted that nuclear mutations were likely to be responsible for those defects which had tissue specificity or were developmentally regulated. Since then, the pendulum has swung away from this view and variation in level of heteroplasmy has been used to explain these features. However, there are defects in respiratory function which are clearly autosomally inherited, and it is likely that variability in diseases attributed to mitochondrial DNA mutations may involve interactions with nuclear isoforms. For example, of the 13 or so subunits of cytochrome oxidase, only three are mitochondrially encoded. There is now very good evidence for tissue specific isoforms of subunits VI and VII, and variation in the protein sequence which may be polymorphic in health could interact with certain mitochondrial DNA mutations resulting in tissue specific respiratory defects. Similarly, some of the nuclear encoded subunits of ATP synthase are developmentally regulated, and could explain some of the variation in age of presentation which is observed.

Major Rearrangements of Mitochondrial DNA: Simple Deletions

The first evidence that rearrangements of mitochondrial DNA might cause human disease was obtained by Holt et al. (1988). There were a number of groups aiming to identify mutations in the mitochondrial myopathies at this time, all working on the most accessible tissue, blood. All were aware that heteroplasmy was a possibility, but obtaining muscle from these rare patients was difficult: molecular biologists were competing with biochemists bent on purifying mitochondria for analysis of the respiratory chain from milligram quantities of tissue. Ian Holt realized that while he would not be able to analyze mitochondrial DNA from these precious extracts, there was a residual pellet containing nuclei and other cell debris, normally discarded, in which mitochondrial DNA was a significant contaminant. Southern[20] hybridization (see Chapter 3) demonstrated large deletions in this mitochondrial

DNA from muscle, which were not detectable in blood. These were found in patients with KSS or CPEO.

Evidence is accumulating that the deletions cause both KSS and CPEO, but there are still aspects of the pathophysiology which are poorly understood. We will first describe the rearrangements which have been observed and then discuss the experimental data with respect to some key questions, including the relationship of the deletions to the phenotype.

The "Common Deletion"

All the deletions described so far (Schon et al., 1989; Holt et al., 1989a; Mita et al., 1990) include one or more tRNA genes and part or all of one or more protein reading frames. The most frequent deletion (Schon et al., 1989), the so-called "common deletion" (Figure 2B), is about 4.9 kb in length and is found in approximately 50% of these patients (Holt et al., 1989b). Part or all of seven protein reading frames are deleted (see Figure 2B). How do these rearrangements match up to the clinical features? Surprisingly, Moraes et al. (1989) found no clear relationship of the phenotype either to the region of the mitochondrial genome deleted, the length of mitochondrial DNA deleted (1 to 7 kb), or the percentage of mitochondrial DNAs which are deleted (which may range from 20 to 80% in muscle). The identical deletion may give rise to varied phenotypes (Rotig et al., 1988; Schon et al., 1989), mild or severe. Where a number of biopsies have been taken from the same individual, the proportion of deleted genomes increases with time (Larsson et al., 1990), consistent with clinical deterioration. Where multiple biopsies have been taken from different sites in the same individual they are not significantly different. Thus, there may be a dosage effect: clinical severity increases with increasing proportion of deleted genomes within an individual or tissue.

Does the sequence data give us any clues as to how these rearrangements occurred? Recombination is not a feature of mitochondrial DNA in higher organisms (see above). Sequence analysis of the so-called "common deletion" (Schon et al., 1989) reveals that there are direct repeats at either end of the deleted segment, 13 bp in length. Furthermore, it appears that this particular stretch may be a recombination "hot spot" as a number of deletions share one end or the other. Sequence data from an increasingly large number of deletions shows that many but not all have similar direct repeats flanking the deleted segment, suggesting that mispairing may be a feature of the recombination event (Mita et al., 1990). For instance, during replication the heavy strand might be particularly vulnerable as it remains single stranded, while the replication fork proceeds 2/3 of the way around the genome.

Many investigators consider that the gold standard for establishing the credentials of a mitochondrial DNA disease is the use of ρ_0(mitochondrial DNA free)-cell lines. If a cell line from a patient has a detectable biochemical defect which can be transferred to these lines by making cytoplasts and fusing them with the ρ_0-line to form cybrids, then the biochemical defect must be encoded by mitochondrial DNA. If not, the defect resides in the nucleus. Hayashi et al. (1994b) demonstrated that all mitochondrial protein synthesis was impaired in cybrids containing high levels of mitochondrial DNA deletions, presumably because the mitochondria were unable to translate any mRNA without a full complement of 22 tRNAs. When the proportion of mutant was below 60%, the proteins synthesized included an abnormal polypeptide corresponding to the fusion gene formed by the deletion. Above this level, a severe translational defect was apparent. They suggested that there must be free communication between mitochondria so that wild-type genomes "complemented" the mutants. Thus it appeared that the deletions were pathogenic because there was a dose relationship between mitochondrial DNA level and severity of the respiratory defect, and because the mitochondrial DNA mutations were confined to patients with a distinctive phenotype.

Duplications of Mitochondrial DNA

This was taken a stage further by identifying partial duplications of mitochondrial DNA about 8 kb in length in some patients with deletions. Each duplicated molecule had two copies of the D-loop, and hence four origins of replication. Three cases with duplications of mitochondrial DNA demonstrated that there were, in addition, two forms of closed circular deletions, namely a deletion monomer and a dimer. Lower levels of duplications of mitochondrial DNA were detectable in almost all of the patients with KSS, suggesting that they are a hallmark of KSS.

Because there was never more than one abnormal junction fragment on any restriction digest, these three types of mutant mitochondrial DNA are closely related and probably derived from a single illegitimate recombination event, followed by resolution with wild-type. It was suggested that some or all of the deletions in muscle derive from duplications for the following reasons. First, there are several reports of familial mitochondrial DNA duplications (Dunbar et al., 1993), while the majority of pure deletions appear sporadic. Second, deletions appear to accumulate in muscle while duplications decrease. It may be that duplications allow a wider distribution of rearranged mitochondrial DNA molecules in KSS than CPEO (Zeviani et al., 1990), consistent with the multi-system clinical involvement.

There are no known parallels for these novel and complex interactions between families of recombinant mitochondrial and nuclear genomes in closely related organisms, although similar events do occur in plants (see above).

Pearson's Syndrome

Deletions which are identical to those identified in KSS may be found in Pearson's syndrome. This is characterized by sideroblastic anemia, often accompanied by a deficiency of neutrophils and platelets, pancreatic dysfunction, and abnormal liver function, but no neurological symptoms (Pearson et al., 1979). Identical deletions may be found in both of these apparently different phenotypes

(Rotig et al., 1988; Schon et al., 1989) However, the tissue distribution of mutants is very different, being high in blood in Pearson's syndrome (Rotig et al., 1989, 1990), but localized to muscle (Moraes et al., 1989) and the central nervous system in KSS. The hematological abnormalities in Pearson's syndrome may resolve, and the phenotype evolve into KSS (Larsson et al., 1990; McShane et al., 1991; Baerlocher et al., 1992), and it would appear that this is caused by a change in the distribution of mutants: they appear to be lost from blood and accumulate in muscle.

Point Mutations of Mitochondrial DNA

Point mutations of mitochondrial DNA have been described in: Leber's Hereditary Optic Neuropathy (LHON); two mitochondrial myopathies mercifully abbreviated "MELAS" (Goto et al., 1990) and "MERRF" (Shoffner et al., 1990) ("mitochondrial encephalomyopathy, lactic acidosis, and strokes"; and "mitochondrial encephalomyopathy and ragged red fibers," respectively); Leigh's encephalopathy (Tatuch et al., 1992); and a cardiomyopathy (Zeviani et al., 1991). While the association of mutation and phenotype is excellent in most of these, a number of other mutations have been described where the association with abnormal phenotype is less clearly defined (Johns and Berman, 1991; Lauber et al., 1991). Because mitochondrial DNA is highly polymorphic, one would expect that there would be many differences from the reference sequence in any complete mitochondrial DNA sequenced. After excluding those which do not give rise to an amino acid substitution, there may be several candidates for the causative mutation. In the absence of a clear biochemical defect, their significance is extremely hard to investigate. Three examples of point mutations will be discussed briefly.

Leber's Hereditary Optic Neuropathy (LHON)

Patients with LHON show bilateral blindness as adolescents or young adults. Although strictly maternally inherited, the incidence is much higher in males than females, suggesting that nuclear gene products must also be involved in the phenotype. Three point mutations have now been closely associated with the phenotype at bp 11,778 (Wallace et al., 1988), at bp 3460 (Howell et al., 1991; Huoponen et al., 1991), and at bp 14,484 (Johns et al., 1993). Each mutation results in substitution of a conserved amino acid and would be expected to alter protein structure and function. Unlike many other mitochondrial DNA diseases, homoplasmy for the mutation is frequently found but penetrance is incomplete and sex-related. The phenotype may depend on a complex interaction between several mitochondrial and nuclear gene products. The absence of a clear biochemical defect in ND4 makes this problem particularly hard to investigate.

Mitochondrial DNA in MERRF

A point mutation in lysyl tRNA has been found in the majority of patients with MERRF (Shoffner et al., 1990). This lies within the psi loop of the tRNA and would be expected to "bend" the tRNA and might result in faulty lysine incorporation. Translation products with the highest number of lysine codons are reduced, suggesting that this hypothesis is correct (Chomyn et al., 1991). Similarly, Chomyn recently found that polysomes were smaller in cell lines with the translational defect (Enriquez et al., 1995), suggesting stalling of the growing amino acid chain and release of ribosomes. This is supported by identification of several additional peptides which are probably the most stable of the corresponding degradation products.

Mitochondrial DNA in MELAS

MELAS is associated with several point mutations including two in leucyl tRNA at positions 3243 and 3271. Unlike MERRF, pulse labeling of translation products does not clearly implicate faulty amino acid incorporation. As well as possibly influencing tRNA function, the 3243 mutation lies within the binding site for the so-called termination factor involved in RNA processing (Hess et al., 1991). *In vitro*, the point mutation reduces binding of the termination factor and results in a lower level of the shorter of two alternate transcripts. However, this is not reflected by different levels *in vivo*, where the main abnormal finding with both mutations is of a slightly increased level of a transcript known as RNA 19, comprising the 16S ribosomal RNA-tRNA^{leu}-ND1 polycistronic transcript. The mechanisms producing these defects are still uncertain.

Perhaps the most perplexing thing about the 3243 mutation is the wide range of phenotypes that it appears to cause. In MELAS, the patients may come to medical attention in childhood or early teenage because they are small, developmentally delayed, and are having stroke-like episodes and/or seizures. Alternatively, they can present with CPEO, diabetes, deafness, cardiomyopathy, or any combination. This variability is unlikely to be due to variation in "dose" of the mutant alone. Other possible influences include intracellular distribution of mutants and interacting mitochondrial and/or nuclear genes.

SENESCENCE

Fungi

The fungus *Podospora anserina* cannot be cultured vegetatively for long periods. It stops growing and eventually dies. This process of senescence is akin to aging in other organisms. Orgel (1963) suggested that cells age due to the accumulation of harmful mutations. In *Podospora anserina* the senescent state is maternally inherited and can be transmitted vegetatively. In senescent mycelia the first intron of the cytochrome oxidase subunit I (*cox* I) gene is amplified as a separate DNA molecule called sen DNAa. Excision of sen DNAa appears to mediate senescence either by altering the regulation of the *cox*1 gene or perhaps by the lack of production of a maturase. The maturase, which may be needed to excise other mitochondrial introns, can only be produced if the intron remains fused to exon I of the *cox* 1 gene and will not be produced after excision (Benne and Tabak, 1986).

Man

Is mitochondrial DNA also involved in human aging? There is good evidence for a slight decline in the activity of the respiratory chain with aging, along with the accumulation of low levels of several mitochondrial DNA deletions (in most studies <1% total mitochondrial DNA, but up to a remarkable 30% in one case). However, many other changes occur in aging cells: telomeres shorten, there is free radical damage to DNA and proteins, etc. Aging in fibroblasts showed a slight decline in respiratory function in late passages (Hayashi et al., 1994a), but no respiratory deficit was detected when cybrids were formed by fusion of cytoplasts from aging fibroblasts with ρ_0-cells. The respiratory defect could be complemented by fusing with younger fibroblasts, and the authors concluded that the phenotype is nuclear in origin. However, it has been argued that aging in cell lines is not relevant to aging in the nondividing cells of aging humans. Attardi's group has recently demonstrated a cytoplasmic component which correspondingly declines in respiratory function with age. Doubtless the controversy will continue for several more rounds of new mitochondrial DNA deletions in our brains.

NEW METHODS FOR STUDYING ORGANELLE GENOMES

In 1988, Boynton and co-workers reported successful transformation of *C. reinhardtii* plastids with plasmid DNA. At the same time successful transformation of yeast mitochondria was also reported (Johnston et al., 1988). These successes were due to the development of a novel method for introducing DNA into cells (see Chapter 3). The particle bombardment method was developed by a team at Cornell University and involved bombarding cells with high velocity microprojectiles coated with DNA (Klein et al., 1987). The particles penetrate the organelles and deliver multiple copies of plasmid DNA. The use of *C. reinhardtii* allowed powerful selection methods for obtaining the relatively few transformants (around 1–10) among the million recipient cells. The recipient cells contained a deletion in their plastid genes coding for the β-subunit of ATP synthase and did not photosynthesize. Introduction of the wild-type *atpB* gene on a plasmid replaced the deleted gene by homologous recombination in regions flanking the lesion. The resulting repaired plastid genome allowed transformants to photosynthesize and grow into small green colonies in the presence of light and atmospheric carbon-dioxide.

The *C. reinhardtii* plastid occupies as much as 40% of the cell volume and has a cross-sectional diameter of 5–10 μm. It is a good target. Yeast mitochondria have a mean diameter of 0.3–0.4 μm and occupy around 10% or less of the cell volume. Mitochondrial transformation in yeast is less efficient than plastid transformation in *C. reinhardtii*. Plastid transformation in *C. reinhardtii* has developed into a fine art since it was first reported in 1988. The development of an expression cassette that mediates spectinomycin resistance allowed reverse genetics to be applied to plastids (Goldschmidt-Clermont, 1991). Plastid genes of unknown function could be mutated by the insertion of the cassette and the consequences of "gene knock-outs" followed.

Transformation opens up a route towards engineering organelles. Endogenous genes can be altered and novel genes introduced into organelles.

ORGANELLE INHERITANCE

In many organisms both plastids and mitochondria are inherited from the female parent. This observation provided one of the first pieces of evidence that led to the discovery of extranuclear DNA (see above). In some cases maternal inheritance of organelles can be explained by the larger cytoplasmic contribution of the female gamete to the zygote. For example, in woman the egg contributes most of the cytoplasm to the zygote. Very little of the sperm cytoplasm accompanies the nucleus on fertilization and the frequency of transmission of paternal mitochondria is low—0.01–0.1% of the total (Gyllenstein et al., 1990).

Hauswirth and Laipis investigated a herd of Holstein cows for a common polymorphism (Laipis et al., 1988). They found that when there is a point mutation difference between a mother and her offspring, there is usually a complete switching in a single generation: each is homoplasmic with regard to that point mutation. Because oocytes contain tens of thousands of mitochondrial DNAs and yet the mutation probably only occurred once, they suggested that only a small number of mitochondrial DNAs were selected to populate the organism (the so-called "bottle-neck" hypothesis). By analysis of the individuals where the switching was not complete, they were able to estimate that this number of mitochondrial DNAs was between 1 and 6. When they repeated the experiment in mice they reached a slightly higher estimate. Thus all of the millions of mitochondrial DNAs in an individual may differ by a single bp from all of the mitochondrial DNAs in their mother. There is, however, no known mechanism for this bottleneck. One could perhaps imagine a nuclear "switch" gene which made a single product which "switched on" replication in a single mitochondrion. However, the progeny would need to be similarly "switched on" or the remainder "switched off," and it is difficult to hypothesize a mechanism whereby this could be established.

In *C. reinhardtii* plastid DNA is normally only inherited from the mating type plus parent. The equal cell size of both gametes means that differences in cytoplasmic volume cannot explain uniparental inheritance of organelles. Indeed, mitochondrial DNA in contrast to plastid DNA is inherited from the mating type minus parent (Harris, 1989). Sager has suggested that plastid DNA in the mating type minus parent is degraded by nucleases that cannot act on the plus parent whose plastid genome is methylated (Sager, 1985). The characterization of a nuclear gene affecting plastid DNA transmission will improve our understanding of uniparental inheritance in *C. reinhardtii* (Armbrust et al., 1993).

A number of mechanisms have been proposed to explain maternal inheritance of plastids in higher plants (Hagemann and Schröder, 1989). These include the elimination or degradation of plastids from the male sperm nuclei. Cereal plants regenerated from pollen often contain white plastids with deleted genomes (Day and Ellis, 1984). These could arise as a consequence of processes that debilitate plastids during pollen maturation so that functional plastids are only derived from the egg cell. Maternal inheritance of plastids is not universal in plants. In angiosperms biparental inheritance of plastids is well documented and paternal inheritance has been reported in some species including carrots (Boblenz et al., 1990). Paternal inheritance of plastids is common in gymnosperms. It has been argued that uniparental inheritance of organelles is an inevitable consequence of evolutionary pressures that reduce the spread of selfish cytoplasmic elements (Law and Hutson, 1992).

IS EXTRANUCLEAR DNA LOCATED OUTSIDE MITOCHONDRIA AND PLASTIDS?

The analysis of non-Mendelian inheritance in the early part of this century led to the discovery of DNA in mitochondria and plastids. However, not all instances of non-Mendelian inheritance are due to extranuclear DNA. For example, lack of transmission of paternal genes can result from apomixis—a process by which plants develop from seeds without sexual fertilization. In some cases DNA located in the nucleus is responsible for non-Mendelian segregation patterns. Multicopy nuclear plasmids in fungi exhibit non-Mendelian inheritance because they are extra-chromosomal. DNA due to infectious agents such as viruses can be found outside the nucleus, but since they are neither universal nor permanent inhabitants of a cell type they have not been discussed. Are there any other organelles besides nuclei, mitochondria, and plastids that might contain DNA?

Basal bodies found at the base of flagella are proposed to be the remnants of a once free-living spirochaete-like bacterium (reviewed in Sagan, 1967). This endosymbiotic event conferred motility on the recipient protoeukaryotic cell. A recent claim that the basal body of *Chlamydomonas reinhardtii* contains DNA (Hall et al., 1989) has not been substantiated (Johnson and Rosenbaum, 1990).

Lessons from mitochondria and plastids suggest that at least two criteria govern the presence of organellar DNA. First, the derivation of organelles from endosymbionts, and second, the retention of some of the original endosymbiont's genes by the organelle. If all the genes in the original endosymbiont were relocated to the nucleus, traces of their ancestry would be lost. The retention of DNA in mitochondria and plastids provided the basis for a rich scientific adventure that has increased our understanding of the origin of eukaryotic cells, and expanded our views on inheritance, gene expression, and genome evolution.

ACKNOWLEDGMENTS

We would like to thank Professors Reinhold Herrmann and Masahiro Sugiura and Springer-Verlag for their kind permission to reproduce Figures 1 and 3; Professor Giorgio Bernardi for assistance in finding a picture of Professor Ephrussi, the Royal Society and the Biotechnology and Biological Sciences Research Council (AD) and Wellcome Trust (JP), for financial support. We are grateful to very many colleagues for stimulating discussions, in particular Garry Brown and Karl Morten.

REFERENCES

Anderson. S, Bankier, A.T., Barrell, B.G., de Bruijn, M.H.L., Coulson, A.R., Drouin, J., Eperon, I.C., Nierlich, D.P., Roe, B.A., Sanger, F., Schrier, D.H., Smith, A.J.H., Staden, R., & Young, I.G. (1981). Sequence and organization of the human mitochondrial genome. Nature, Lond. 290, 457–465.

Armbrust, E.V., Ferris, P.J., & Goodenough, U.W. (1993). A mating type-linked gene cluster expressed in *C. reinhardtii* zygotes participates in the uniparental inheritance of the chloroplast genome. Cell 74, 801–811.

Attardi, G. (1985). Animal mitochondrial DNA: an extreme example of genetic economy. Int. Rev. Cytol. 93, 93–145.

Attardi, G. & Schatz, G. (1988). Biogenesis of mitochondria. Annu. Rev. Cell Biol. 4, 289–333.

Baldauf, S.L. & Palmer, J.D. (1990). Evolutionary transfer of the chloroplast *tufA* gene to the nucleus. Nature, Lond. 344, 262–265.

Baerlocher, K.E., Feldges, A., Weissert, M., Simonsz, H.J., & Rotig, A. (1992). Mitochondrial DNA deletion in an 8-year-old boy with Pearson syndrome. J. Inherit. Metab. Dis. 15, 327–330.

Baur, E. (1909). Das Wesen und Erblichkeitsverhältnisse der varietates "albomarginatae hort." von *Pelargonium zonale*. Vererbungsl. 1, 330–351.

Bedbrook, J.R. & Bogorad, L. (1976). Endonuclease recognition sites mapped on *Zea mays* chloroplast DNA. Proc. Natl. Acad. Sci. USA 73, 4309–4313.

Belliard, G., Vedel, M., & Pelletier, G. (1979). Mitochondrial recombination in cytoplasmic hybrids of *Nicotiana tabacum* by protoplast fusion. Nature, Lond. 281, 401–403.

Benne, R. (1988). Aminoacyl-tRNA synthetases are involved in RNA splicing in fungal mitochondria. Trends Genet. 4, 181–182.

Benne, R. & Tabak, H.F. (1986). Senescence comes of age. Trends Genet. 2, 147–148.

Bernardi, G. (1979). The *petite* mutation in yeast. Trends Biochem. Sci. 4. 197–201.

Bibb, M.J., Van Etten, R.A., Wright, C.T., Walberg, M.W., & Clayton, D.A. (1981). Sequence and gene organization of mouse mitochondrial DNA. Cell 26, 167–180.

Boblenz, K., Nothnagel, T., & Metzlaff, M. (1990). Paternal inheritance of plastids in *Daucus*. Mol. Gen. Genet. 220, 489–491.

Boer, P.H. & Gray, M.W. (1988). Scrambled ribosomal RNA gene pieces in *Chlamydomonas reinhardtii* mitochondrial DNA. Cell 55, 399–411.

Boogaart, P. van den, Samallo, J., & Agsteribbe, E. (1982). Similar genes for a mitochondrial ATPase subunit in the nuclear and mitochondrial genomes of *Neurospora crassa*. Nature, Lond. 298, 187–189.

Borst, P. & Rutenberg, G.J.C.M. (1966). Renaturation of mitochondrial DNA. Biochim. Biophys. Acta 114, 645–647.

Borst, P. & Grivell, L.A. (1981). Small is beautiful-portrait of a mitochondrial genome. Nature, Lond. 290, 443–444.

Boynton et al. (1988). Chloroplast transformation in *Chlamydomonas* with high velocity microprojectiles. Science 240, 1534–1537.

Britten, R.J. & Kohne, D.E. (1968). Repeated sequences in DNA. Science 161, 529–540.

Brown, W.M. & Vinograd, J. (1974). Restriction endonuclease cleavage maps of animal mitochondrial DNAs. Proc. Natl. Acad. Sci. USA 71, 4617–4621.

Bruggen, E.F.J. van, Borst, P., Rutenberg, G.J.C.M., Gruber, M., & Kroon, A.M. (1966). Circular mitochondrial DNA. Biochim. Biophys. Acta 119, 437–439.

Cann, R.L., Stoneking, M., & Wilson, A.C. (1987). Mitochondrial DNA and human evolution. Nature, Lond. 325, 31–36.

Chang, D.D. & Clayton, D.A. (1987). A mammalian mitochondrial RNA processing activity contains nucleus-encoded RNA. Science 235, 1178–1184.

Clark-Walker, G.D. & Linnane, A.W. (1967). The biogenesis of mitochondria in *Saccharomyces cerevisiae*. J. Cell Biol. 34, 1–14.

Chomyn, A., Meola, G., Bresolin, N., Lai, S.T., Scarlato, G., & Attardi, G. (1991). *In vitro* genetic transfer of protein synthesis and respiration defects to mitochondrial DNA-less cells with myopathy-patient mitochondria. Mol. Cell Biol. 11, 2236–44.

Correns, C. (1900). G. Mendels Regel über da Verhalten der Nachkommenschaft der Rassenbartarde. Ber. dt. bot. Ges 18, 158–168.

Correns, C. (1909). Vererbungsversuche mit blaβ(gelb)grünen und buntblättrigen Sippen bei *Mirabilis jalapa*, *Urtica pilufera* und *Lunaria annua*. Z. Verebungsl. 1, 291–329.

Day, A. & Ellis, T.H.N. (1984). Chloroplast DNA deletions associated with wheat plants regenerated from pollen: possible basis for maternal inheritance of chloroplasts. Cell 39, 359–368.

Dietrich, A., Weil, J.H., & Maréchal-Drouard, L. (1992). Nuclear-encoded transfer RNAs in plant mitochondria. Annu. Rev. Cell Biol. 8, 115–131.

Dujon, B., Belfort, M., Butow, R.A., Jacq, C., Limieux, C., Perlman, P.S., & Vogt, V.M. (1989). Gene 82, 115–118.

Dunbar, D., Moonie, P., Swingler, R., Davidson, D., Roberts, R., & Holt, I. (1993). Maternally transmitted partial direct tandem duplication of mitochondrial DNA associated with diabetes mellitus. Human Molecular Genetics 2, 1619–1624.

Ellis, R.J. (1977). Protein synthesis by isolated chloroplasts. Biochim. Biophys. Acta 463, 185–215.

Ellis, R.J. (1979). The most abundant protein in the world. Trends Biochem. Sci. 4, 241–244.

Ellis, R.J. (1982). Promiscuous DNA chloroplast genes inside plant mitochondria. Nature, Lond. 299, 678–679.

Ellis, R.J. (1987). Proteins as molecular chaperones. Nature, Lond. 328, 378–379.

Ellis, R.J. & van der Vies, S.M. (1991). Molecular chaperones. Annu. Rev. Biochem. 60, 321–347.

Enriquez, J., Chomyn, A., & Attardi, G. (1995). MtDNA mutation in MERRF syndrome causes defective aminoacetylation of tRNALys and premature translation termination. Nature Genetics 10, 47–55.

Ephrussi, B. (1949). In: Unites Biologiques Douees de Continuite Genetique, Paris, Juin-Juillet, 1948, pp. 165–180. CNRS, Paris.

Farrelly, F. & Butow, R.A. (1983). Rearranged mitochondrial genes in the yeast nuclear genome. Nature, Lond. 301, 296–301.

Forde, B.G., Oliver, R.J.C., & Leaver, C.J. (1978). Variation in mitochondrial translation products associated with male-sterile cytoplasms in maize. Proc. Natl. Acad. Sci. USA 75, 3841–3845.

Fox, T.D. (1986). Nuclear gene products required for translation of specific mitochondrially coded mRNAs in yeast. Trends Genet. 2, 97–100.

Gatenby, A.A., Castleton, J.A., & Saul, M.W. (1981). Expression in *E. coli* of the maize and wheat chloroplast genes for the large subunit of ribulose bisphosphate carboxylase. Nature, Lond. 291, 117–121.

Giles, R.E., Blanc, H., Cann, H.M., & Wallace, D.C. (1980). Maternal inheritance of human mitochondrial DNA. Proc. Natl. Acad. Sci. USA 77, 6715–6719.

Goldschmidt-Clermont, M. (1991). Transgenic expression of aminoglycoside adenine transferase in the chloroplast: a selectable marker for site-directed transformation of *Chlamydomonas*. Nucleic Acids Res. 19, 4083–4089.

Golenberg, E.M., Giannasi, D.E., Clegg, M.T., Smiley, C.J., Durbin, M., Henderson, D., & Zurawski, G. (1990). Chloroplast DNA from a miocene Magnolia species. Nature, Lond. 344, 656–658.

Goto, Y-I., Nonaka, I., & Horai, S. (1990). A mutation in the tRNA leu(UUR) gene associated with the MELAS subgroup of mitochondrial encephalomyopathies. Nature, Lond. 348, 651–653.

Gray, M.W. (1993). Origin and evolution of organelle genomes. Curr. Opin. Genet. Dev. 3, 884–890.

Grivell, L. (1983). Mitochondrial DNA. Sci. Am. 248, 78–89.

Gyllensten, U., Wharton, D., Josefsson, A., & Wilson, A.C. (1991). Paternal inheritance of mitochondrial DNA in mice. Nature, Lond. 352, 255–257.

Hagemann, R. & Schröder, M-B. (1989). The cytological basis of plastid inheritance in angiosperms. Protoplasma 152, 57–64.

Hall, J.H., Ramanis, Z., & Luck, D.J.L. (1989). Basal body/centriolar DNA: molecular genetic studies in *Chlamydomonas*. Cell 59, 121–132.

Harris, E.H. (1989). The *Chlamydomonas* Sourcebook. Academic Press, London.

Hayashi, J. et al. (1994a). Nuclear but not mitochondrial genome involvement in human age-related mitochondrial dysfunction. Functional integrity of mitochondrial DNA from aged subjects. J. Biol. Chem. 269, 6878–83.

Hayashi, J., Takemitsu, M., Goto, Y., & Nonaka, I. (1994b). Human mitochondria and mitochondrial genome function as a single dynamic cellular unit. J. Cell Biol. 125, 43–50.

Hess, J.F., Parisi, M.A., Bennett, J.L., & Clayton, D.A. (1991). Impairment of mitochondrial transcription termination by a point mutation associated with the MELAS subgroup of mitochondrial encephalomyopathies. Nature, Lond. 351, 236–239.

Hiratsuka et al. (1989). The complete sequence of the rice (*Oryza sativa*) chloroplast genome: intermolecular recombination between distinct tRNA genes accounts for a major plastid DNA inversion during the evolution of the cereal. Mol. Gen. Genet. 217, 185–194.

Hoch, B., Maier, R.M., Appel, K., Igloi, G.L., & Kossel, H. (1991). Editing of a chloroplast mRNA by creation of an initiation codon. Nature, Lond. 353, 178–180.

Holt, I.J., Harding, A.E., Cooper, J.M., Shapira, A.H.V., Toscano, A., Clarke, J.B., & Morgan-Hughes, J.A. (1989a). Mitochondrial myopathy: clinical and biochemical features of 30 patients with major deletions of muscle mitochondrial DNA. Ann. Neurol. 26, 699–708.

Holt, I.J., Harding, A.E., & Morgan-Hughes, J.A. (1988). Deletions in muscle mitochondrial DNA in patients with mitochondrial myopathies. Nature, Lond. 331, 717–719.

Holt, I.J., Harding, A.E., & Morgan-Hughes, J.A. (1989b). Deletions of muscle mitochondrial DNA in mitochondrial myopathies: sequence analysis and possible mechanisms. Nucleic Acids Res. 17, 4465–4469.

Howell, N., Bindoff, L.A., McCullough, D.A., Kubacka, I., Poulton, J., Mackey, D., Taylor, L., et al. (1991). Leber hereditary optic neuropathy: identification of the same mitochondrial ND1 mutation in six pedigrees. Am. J. Hum. Genet. 49, 939–50.

Huoponen, K., Vilkki, J., Aula, P., Nikoskelainen, E.K., & Savontaus, M.L. (1991). A new mitochondrial DNA mutation associated with Leber hereditary optic neuroretinopathy. Am. J. Hum. Genet. 48, 1147–53.

Johns, D.R. & Berman, J. (1991). Alternative, simultaneous complex I mitochondrial DNA mutations in Leber's hereditary optic neuropathy. Biochem. Biophys. Res. Commun. 174, 1324–30.

Johns, D.R., Heher, K.L., Miller, N.R., & Smith, K.H. (1993). Leber's hereditary optic neuropathy. Clinical manifestations of the 14484 mutation. Arch. Ophthalmol. 111, 495–498.

Johnson, K.A. & Rosenbaum, J.L. (1990). The basal bodies of *Chlamydomonas reinhardtii* do not contain immunologically detectable DNA. Cell 62, 615–619.

Johnston, S.A., Anziano, P.Q., Shark, K., Sanford, J.C., & Butow, R.A. (1988). Mitochondrial transformation in yeast by bombardment with microprojectiles. Science 240, 1538–1541.

Kirk, J.T.O. & Tilney-Bassett, R.A.E. (1967). The Plastids, 1st ed. W.H. Freeman, London.

Klein, T.M., Wolf, E.D., Wu, R., & Sanford, J.C. (1987). High velocity microprojectiles for delivering nucleic acids into living cells. Nature 327, 70–73.

Kleinschmidt, A.K. (1968). Monolayer techniques in electron microscopy of nucleic acid molecules. Methods Enzymol. 12B, 361–379.

Kolodner, R. & Tewari, K.K. (1975). The molecular size and conformation of the chloroplast DNA of higher plants. Biochim. Biophys. Acta 402, 372–390.

Kutzelnigg, H. & Stubbe, W. (1974). Investigations on plastome mutants in *Oenothera*. Sub. Cell Biochem. 3, 73–89.

Laipis, P.J., Van der Walle, M.J., & Hauswirth, W.W. (1988). Unequal partitioning of bovine and mitochondrial genotypes among siblings. Proc. Natl. Acad. Sci. USA 85, 8107–8110.

Larsson, N.G., Holme, E., Kristiansson, B., Oldfors, A., & Tulinius, M. (1990). Progressive increase of the mutated mitochondrial DNA fraction in Kearns-Sayre syndrome. Pediatr. Res. 28, 131–136.

Laskey, R.A., Honda, B.M., Mills, A.D., & Finch, A.T. (1978). Nucleosomes are assembled by an acidic protein which binds histones and transfers them to DNA. Nature, Lond. 275, 416–420.

Lauber, J., Marsac, C., Kadenbach, B., & Seibel, P. (1991). Mutations in mitochondrial tRNA genes: a frequent cause of neuromuscular diseases. Nucleic Acids Res. 19, 1393–1397.

Law, R. & Hutson, V. (1992). Intracellular symbionts and the evolution of uniparental cytoplasmic inheritance. Proc. Roy. Soc. Lond. B. 248, 69–77.

Lazowska, J., Jacq, C., & Slonimski, P.P. (1980). Sequence of introns and flanking exons in wildtype and BOX3 mutants of cytochromne B reveals an interlaced splicing protein coded by an intron. Cell 22, 333–349.

Leblanc, C., Boyer, C., Richard, O., Bonard, G., Grienberger, J.M., & Kloareg, B. (1995). Complete sequence of mitochondrial DNA of the rodophyte *Chondrus crispus* (Gigartinales). Gene content and genome organization. J. Mol. Biol. 250, 484–495.

Levings, C.S. & Brown, G.G. (1989). Molecular biology of plant mitochondria. Cell 56, 171–179.

Lindegren, C.C. (1949). The Yeast Cell, Its Genetics and Cytology, 1st edn. Educational Publishers, Saint Louis.

Liu, X.-Q., Gillham, N.W., & Boynton, J.E. (1989). Chloroplast ribosomal protein gene *rps12* of *Chlamydomonas reinhardtii*. Wild-type sequence, mutation to streptomycin resistance and dependence and function in *Escherichia coli*. J. Biol. Chem. 264, 16100–16108.

Lonsdale, D.M. (1986). Viral RNA in mitochondria. Nature, Lond. 323, 399.

Lyttleton, J.W. (1962). Isolation of ribosomes from spinach chloroplasts. Exp. Cell Res. 26, 312–317.

McShane, M.A., Hammans, S.R., Sweeney, M., Holt, I.J., Beattie, T.J., Brett, E.M., & Harding, A.E. (1991). Pearson syndrome and mitochondrial encephalomyopathy in a patient with a deletion of mtDNA. Am. J. Hum. Genet. 48, 39–42.

Michaelis, G., Vahrenholz, C., & Pratje, E. (1990). Mitochondrial DNA of *Chlamydomonas reinhardtii*. Mol. Gen. Genet. 223, 211–216.

Michel, F. & Dujon, B. (1983). Conservation of RNA secondary structures in two intron families including mitochondrial-, chloroplast- and nuclear-encoded members. EMBO J. 2, 33–38.

Mita, S. et al. (1990). Recombination via flanking direct repeats is a major cause of large-scale deletions of human mitochondrial DNA. Nucleic Acids Res. 18, 561–567.

Moraes, C.T. et al. (1989). Mitochondrial DNA deletions in progressive external ophthalmoplegia and Kearns-Sayre syndrome. N. Engl. J. Med. 320, 1293–1299.

Moraes, C.T. et al. (1991). mtDNA depletion with variable tissue expression: a novel genetic abnormality in mitochondrial diseases. Am. J. Hum. Genet. 48, 492–501.

Nakazono, M. & Hirai, A. (1993). Identification of the entire set of transferred chloroplast DNA sequences in the mitochondrial genome of rice. Mol. Gen. Genet. 236, 341–346.

Narang, F.E., Bell, J.B., Stohl, L.L., & Lambowitz, A.M. (1984). The DNA sequence and genetic organization of a Neurospora mitochondrial plasmid suggests a relationship to introns and mobile elements. Cell 38, 441–453.

Nass, S. & Nass, M.M.K. (1963). Intramitochondrial fibers with DNA characteristics. II. Enzymatic and other hydrolytic treatments. J. Cell Biol. 19, 613–629.

Neupert, W. & Schatz, G. (1981). How proteins are transported into mitochondria. Trends Biochem. Sci. 6, 1–4.

Oda, K., Yamato, K., Ohta, E., Nakamura, Y., Takemura, M., Nozato, N., Akashi, K., Kanegae, T., Ogura, Y., Kohchi, T., & Ohyama, K. (1992). Gene organization deduced from the complete sequence of liverwort *Marchantia polymorpha* mitochondrial DNA. J. Mol. Biol. 223, 1–7.

O'Farrell, P.H. (1975). High resolution two-dimensional electrophoresis of proteins. J. Biol. Chem. 250, 4007–4021.

Orgel, L.E. (1963). The maintenance of the accuracy of protein synthesis and its relevance to ageing. Proc. Natl. Acad. Sci. USA 49, 517–521.

Palmer, J.D. (1985). Comparative organization of chloroplast genomes. Annu. Rev. Genet. 19, 325–354.

Palmer, J.D. (1990). Contrasting modes and tempos of genome evolution in land plant organelles. Trends Genet. 6, 115–120.

Palmer, J.D. (1993). A genetic rainbow of plastids. Nature, Lond. 364, 762–763.

Pamphillis, C.W. de & Palmer, J.D. (1990). Loss of photosynthetic and chlororespiratory genes from the plastid genome of a parasitic flowering plant. Nature, Lond. 348, 337–339.

Parisi, M.A., Xu, B., & Clayton, D.A. (1993). A human mitochondrial transcriptional activator can functionally replace a yeast mitochondrial HMG-box protein both *in vivo* and *in vitro*. Mol. Cell Biol. 13, 1951–61.

Pearson, H. et al. (1979). A new syndrome of refractory sideroblastic anaemia with vacuolisation of marrow precursors and exocrine pancreatic dysfunction. J. Pediatrics 95, 976–984.

Poulton, J. (1987). "All about Eve." New Scientist 1560, 51–53.

Ris, H. & Plaut, W. (1962). Ultrastructure of DNA containing areas in the chloroplast of *Chlamydomonas*. J. Cell Biol. 13, 383–391.

Rochaix, J-D. (1992). Post-transcriptional steps in the expression of chloroplast genes. Annu. Rev. Cell Biol. 8, 1–28.

Rotig, A. et al. (1988). Deletions of blood mitochondrial DNA in pancytopenia. Lancet i, 567–568.

Rotig, A., Colonna, M., Bonnefont, J.P., Blanche, S., Fischer, A., Saudubray, J.M., & Munnich, A. (1989). Mitochondrial DNA deletion in Pearson's marrow/pancreas syndrome [letter; comment] [see comments]. Lancet 1, 902–3.

Rotig, A. et al. (1990). Pearson's marrow-pancreas syndrome, a multisystem mitochondrial disorder in infancy. J. Clin. Invest. 86, 1601–1608.

Sagan, L. (1967). On the origin of mitosing cells. J. Theor. Biol. 14, 225–274.

Sager, R. (1954). Mendelian and non-Mendelian inheritance of streptomycin resistance in *Chlamydomonas*. Proc. Natl. Acad. Sci. USA 40, 356–363.

Sager, R. (1985). Chloroplast genetics. Bioessays 3, 180–184.

Sager, R. & Ishida, M.R. (1963). Chloroplast DNA in *Chlamydomonas*. Proc. Natl. Acad. Sci. USA 50, 725–730.

Schatz, G. (1968). Impaired binding of mitochondrial adenosine triphosphatase in the cytoplasmic "petite" mutant of *Saccharomyces cerevisiae*. J. Biol. Chem. 243, 2192–2199.

Schatz, G. & Mason, T.C. (1974). The biosynthesis of mitochondrial proteins. Annu. Rev. Biochem. 43, 840–887.

Schön, A., Krupp, G., Gough, S., Berry-Lowe, S., Kannangara, G., & Söll, D. (1986). The RNA required in the first step of chlorophyll biosynthesis is a chloroplast glutamate tRNA. Nature, Lond. 322, 281–284.

Schon, E.A., Rizzuto, R., Moraes, C.T., Nakase, H., Zeviani, M., & DiMauro, S. (1989). A direct repeat is a hotspot for large-scale deletion of human mitochondrial DNA. Science 244, 346–349.

Schwarz, Z. & Kössel, H. (1980). The primary structure of 16S rDNA from *Zea mays* chloroplast is homologous to *E. coli* 16S rDNA. Nature, Lond. 283, 739–742.

Seyer, P., Kowallik, K.V., & Herrmann, R.G. (1981). A physical map of *Nicotiana tabacum* plastid DNA including the location of structural genes for ribosomal RNAs and the large subunit of ribulose-bisphosphate carboxylase/oxygenase. Current Genetics 3, 189–204.

Sharp, P.A. (1991). "Five easy pieces." Science 254, 663.

Shih, M-C., Lazr, G., & Goodman, H.M. (1986). Evidence in favour of the symbiotic origin of chloroplasts: primary structure and evolution of tobacco glyceraldehyde-3-phosphate dehydrogenases. Cell 47, 73–80.

Shimada, H. & Sugiura, M. (1991). Fine structural features if the chloroplast genome: comparison of the sequenced chloroplast genomes. Nucleic Acids Res. 19, 983–995.

Shoffner, J.M., Lott, M.T., Lezza, A.M., Seibel, P., Ballinger, S.W., & Wallace, D.C. (1990). Myoclonic epilepsy and ragged-red fiber disease (MERRF) is associated with a mitochondrial DNA tRNA(Lys) mutation. Cell 61, 931–937.

Shyjan, A.W. & Butow, R.A. (1993). Intracellular dialogue. Current Biology 3, 398–400.

Smith, C.A., Jordan, J.M., & Vinograd, J. (1971). *In vivo* effects of intercalating drugs on the superhelix density of mitochondrial DNA isolated from human and mouse cells in culture. J. Mol. Biol. 59, 255–272.

Spreitzer, R.J., Goldschmidt-Clermont, M., Rahire, M., & Rochaix, J-D. (1985). Nonsense mutations in the *Chlamydomonas* chloroplast gene that codes for the large subunit of ribulosebisphosphate carboxylase/oxygenase. Proc. Natl. Acad. Sci. USA 82, 5460–5464.

Stern, D.B. & Lonsdale, D.M. (1982). Mitochondrial and chloroplast genomes of maize have a 12 kb DNA sequence in common. Nature, Lond. 299, 698–702.

Sutton, W.S. (1903). The chromosomes in heredity. Biol. Bull. Mar. Biol. Lab., Woods Hole 4, 231–248.

Tabak, H.F. & Grivell, L.A. (1986). RNA catalysis in the excision of yeast mitochondrial introns. Trends Genet. 2, 51–55.

Tatuch, Y. et al. (1992). Heteroplasmic mtDNA mutation (T-G) at 8993 can cause Leigh disease when the percentage of abnormal mtDNA is high. Am. J. Hum. Genet. 50, 852–959.

Taylor, D.L. (1970). Chloroplasts as symbiotic organelles. Int. Rev. Cytol. 27, 29–64.

Taylor, W.C. (1989). Regulatory interactions between nuclear and plastid genomes. Annu. Rev. Plant Physiol. Plant Mol. Biol. 40, 211–233.

Tewari, K.K. (1979). Structure and replication of chloroplast DNA. In: Nucleic Acids in Plants (Hall, T.C. & Davies, J.W., Eds.), Vol. 1, pp. 41–108. CRC Press.

Thorsness, P.E. & Fox, T.D. (1990). Escape of DNA from mitochondria to the nucleus in *Saccharomyces cerevisiae*. Nature, Lond. 346, 376–379.

Timmis, J.N. & Scott, N.S. (1983). Sequence homology between spinach nuclear and chloroplast genomes. Nature, Lond. 305, 65–67.

Walbot, V. (1991). RNA editing fixes problems in plant mitochondrial transcripts. Trends Genet. 7, 37–39.

Walbot, V. & Coe, E.H. (1979). Nuclear gene *iojap* conditions a programmed change to ribosome-less plastids in *Zea mays*. Proc. Natl. Acad. Sci. USA 76, 2760–2764.

Wallace, D.C. et al. (1988). Mitochondrial DNA mutation associated with Leber's hereditary optic neuropathy. Science 242, 1427–1430.

Weiner, A.M. & Maizels, N. (1990). RNA editing: guided but not templated? Cell 61, 917–920.

Wissinger, B., Brennicke, A., & Schuster, W. (1992). Regenerating good sense. Trends Genet. 8, 322–328.

Wolfe, K.H., Morden, C.W., Ems, S., & Palmer, J.D. (1992). Rapid evolution of the plastid translational apparatus in a non-photosynthetic plant: loss or accelerated sequence evolution of tRNA and ribosomal protein genes. J. Mol. Evol. 35, 304–317.

Yoneda, M., Miyatake, T., & Attardi, G. (1994). Complementation of mutant and wild-type human mitochondrial DNAs coexisting since the mutation event and lack of complementation of DNAs introduced separately into a cell within distinct organelles. Mol. Cell Biol. 14, 2699–712.

Yu, W. & Spreitzer, R.J. (1992). Chloroplast heteroplasmicity is stabilized by an amber-suppressor tryptophan tRNA-CUA. Proc. Natl. Acad. Sci. USA 89, 3904–3907.

Zamaroczy, M. de & Bernardi, G. (1986). The primary structure of the mitochondrial genome of *Saccharomyces cerevisiae*. A review.

Zeviani, M. et al. (1991). Maternally inherited myopathy and cardiomyopathy: association with mutation in mitochondrial DNA tRNA(Leu)(UUR). Lancet 338, 143–7.

Zeviani, M., Gellera, C., Pannacci, M., Uziel, G., Prelle, A., Servidei, S., & DiDonato, S. (1990). Tissue distribution and transmission of mitochondrial DNA deletions in mitochondrial myopathies. Ann. Neurol. 28, 94–7.

GEORGE BEADLE[1]

JEAN BRACHET[2]

EDWARD LAWRIE TATUM[3]

ERWIN CHARGAFF[4]

FRANCIS CRICK[5]

ALFRED HERSHEY[6]

JIM WATSON[7]

ROSALIND FRANKLIN[8]

ARTHUR KORNBERG[9]

ROBERT HOLLEY[10]

SEVERO OCHOA[11]

FRED SANGER[12]

HAR GOBIND KHORANA[13]

HOWARD TEMIN[14]

MARSHALL NIRENBERG[15]

KARY MULLIS[16]

STANLEY COHEN[17]

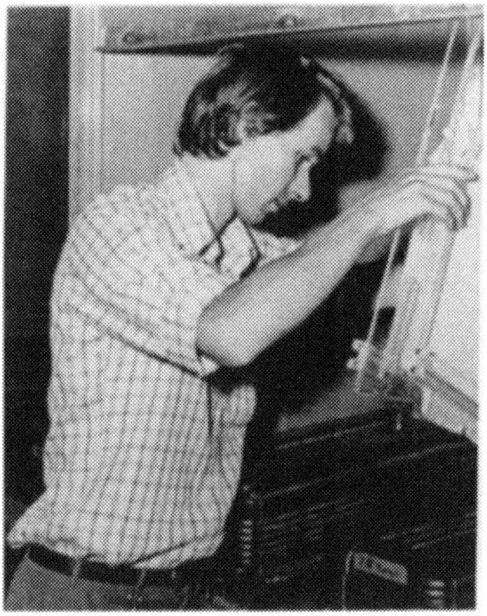

RICH ROBERTS[18]

TOM MANIATIS[19]

ED SOUTHERN[20]

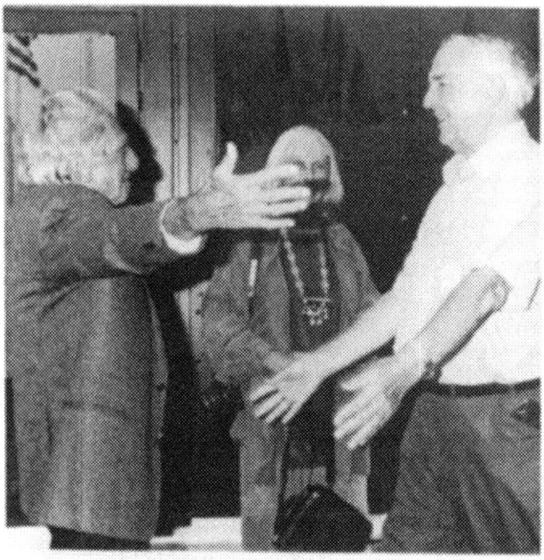

DAN MAZIA[21] and
MURDOCH MITCHISON (r)

WALTER GILBERT[22]

MAX BERGMANN[23]

DON NATHANS[24]

JOSEPH FRUTON[25]

ALBERT CLAUDE[26]

PHILIP SIEKEVITZ[27]

PAUL ZAMECNIK[28]

KEITH PORTER[29]

DOROTHY HODGKIN[30]

GEORGE PALADE[31]

MAX PERUTZ[32]

JOHN KENDREW[33]

BORIS EPHRUSSI[34]

Chapter 5

PROTEIN SYNTHESIS AND THE RIBOSOME

Philip Siekevitz

PROLOGUE

Writing a history of any kind is almost an impossible task, indeed a lost cause in many respects. The beginning is arbitrary, the ending is unknown; conjecture is rife at the beginning and doubt appears at the ending. Scientific history is no exception, even if the historian digs deeply into notebooks, jottings, recorded conversations, archives, anecdotes true or false. As a non-historian, I have not the capacity nor the time to examine all this. So what to do? To do the best one can, to start at an arbitrary point, but to explain this choice, and to end with still many questions unanswered, a requisite for the study of science. In between, one has to choose which references to bring together to form a somewhat coherent story; if some disagree, I hope I will be pardoned. So here goes.

THE BEGINNINGS

The study of protein synthesis really began in the late 1930s and early 1940s with at that time the revolutionary finding that macromolecules are not stable in the body, but undergo constant breakdown and resynthesis. Schoenheimer and Rittenberg, using the newly available deuterium from heavy water, showed that body lipids were undergoing a constant turnover. For us, the relevant experiments were those of Schoenheimer and Ratner who used both administered deuterium and ^{15}N to find that protein actively incorporated these isotopes. Thus came about the concept of the dynamic state of body constituents, as Schoenheimer in 1942 labeled the process. This should not have been too surprising, for earlier Borsook and Keighley had found that nitrogen metabolism in the body was widespread and rapid. But it was really the Schoenheimer book, (1942), *The Dynamic State of Body Constituents*, which caught the attention of biochemists. What to make of all this? The first idea was that peptide bonds were being split and then reconstituted; another idea was that of "wear and tear"—that proteins, through constant usage, "age," become denatured, and are broken down by proteolytic enzymes, to be replaced in due time.

At about that time, knowledge concerning proteolytic enzymes was quite developed, but what about protein synthesis? What was known, from work in the laboratory of Bergmann[23] and Fruton,[25] was that the same enzymes which could break a peptide bond, could also make that same bond under certain conditions of reversibility. The idea that this could be a mechanism for protein synthesis was not that far-fetched, for a key problem was the specificity involved in protein synthesis, and indeed the various proteolytic enzymes were very specific as to the nature of the peptide bonds which were broken. Also, the energy requirement could be met in the peptide bond which was to be split, since there is little energy difference between the bond split and that reformed. It is useful to quote Fruton: "The utilization, by transpeptidation, of the energy available for protein synthesis would require catalysis of extreme specificity which could direct, precisely and reproducibly, the sequence of chemical reactions leading to the formation of a protein. To our knowledge, the proteolytic enzymes are the only available biocatalysts that act on peptide bonds with the requisite specificity."

But it was not to be. As a test, Loftfield and co-workers found in 1953 that radioactive alanine was incorporated into the proteins of liver slices about 100 times more than was the α-aminobutyric acid analogue, but the same slice was equally effective in hydrolyzing alanylglycine and aminobutyrylglycine. The protein synthetic machinery seemed to be much more specific than protein hydrolysis. Thus the energy for peptide bond formation must come from a source other than the splitting of the peptide bond. Indeed, it was already theorized in 1941 by Lipmann and by Kalkar that ATP could be a universal energy source, and that it may phosphorylate amino acids prior to their polymerization into a peptide chain. Based on these ideas, experiments were done by Zamecnik and his colleagues in 1948 to show that oxygen in the liver slice was necessary for radioactive amino acid incorporation into protein, and, more to the point, as found by Frantz et al. in 1948, dinitrophenol, a newly discovered specific inhibitor of mitochondrial ATP formation, also blocked this incorporation. Similarly, many investigators showed that ATP was necessary for the synthesis of simple peptides, such as hippuric acid, pantothenic acid, and glutathione, as mentioned in Borsook's review of 1956.

The really key events in the study of protein synthesis were the inauguration in the 1940s of the availability and use of radioactive tracers, such as ^{15}N, ^{14}C, and ^{35}S, in tracing amino acid and protein metabolism. I mentioned Schoenheimer's use of heavy isotopes, and very rapidly other investigators studied protein turnover *in vivo*, as reviewed by Borsook in 1950. Laboratories such as those of Zamecnik at the Harvard Laboratories at Massachusetts General Hospital in Boston, of Greenberg at the Biochemistry Department at the University of California, Berkeley, and of Borsook at the Californian Institute of Technology, Pasadena, seized on the availability of ^{14}C to synthesize radioactive amino acids, and to use them for studies both *in vivo* and *in vitro*. Excitement was in the air, for it seemed that the end result of every experiment was a revelation. Indeed another fallout from this work was the tracing of the pathways of the intermediary metabolism of amino acids, knowledge of which was scarce at that time. I use the word "incorporation" because in these studies no net protein synthesis ever occurred, nor could it be shown that the labeled amino acid replaced a nonlabeled one, through the breaking and making of the same peptide bond. Thus the neutral term "incorporation" was employed, for, remember, the idea of the linear synthesis of a protein, from one end to another, was not visualized until later.

THE CELL BIOLOGY: EARLY YEARS

It is instructive now to backtrack in time. It had been known by cytologists for years that there existed an area in the cytoplasm which avidly took up basic dyes, was therefore called "basophilic," and was thus undoubtedly acidic in nature. The next advance was the discovery that the basophilia was due to RNA—in 1938 by Brachet[4] using RNAase, in 1939 by Caspersson and Schultz using ultraviolet spectroscopy, and in 1943 by Davidson and Waymouth using chemical methods. A big step, although a speculation from trying to correlate the amount of basophilia, i.e. the amount of RNA, with the purported protein output in the same tissue, was the suggestion by Caspersson in 1941, and Brachet in 1942, that RNA was involved in protein synthesis in the cytoplasm; a heroic advance in thought as it turned out.

At about the same time another avenue of approach began to insert itself into the story. Claude,[26] in trying to isolate the chicken tumor virus by high speed centrifugation, found a fraction, using noninfected chicken cells, which also sedimented, and surprisingly exhibited the same property of containing RNA. His papers, in 1938 and 1941, were the first to show that a nucleic acid component could be isolated from the cytoplasm. In 1943 he gave the name "microsomes" to that fraction which was isolated by high-speed centrifugation, contained RNA, and which later, in 1945, was equated by Porter et al. with a cytoplasmic component as seen with the electron microscope, and which was named "endoplasmic reticulum" by Porter[29] in 1954. Indeed, by 1954 Porter had examined many cell types, seen endoplasmic reticulum in all of them, and begun to equate the reticulum, or "ER," with the basophilia of the earlier investigators. Thus began the process of connecting the ER with protein synthesis.

This work on the ER was continued by Palade[31] in 1956 after he had described, in 1955, small particles on the outer (cytoplasmic) surface of the membranes of the ER. By this time two questions begged to be answered: (1) what was the relationship between the microsome fraction isolated by Claude, which contained RNA, and the basophilia and ER as seen in both light and electron micrographs, and (2) what was the relationship between the above and the small particles on the ER seen by Palade? The answer to these questions came out in two papers in 1956 by Palade and Siekevitz,[27] namely that the microsome fraction was the result of fragmentation of the ER during homogenization and centrifugation. The particles, isolated by detergent treatment which solubilized the membranes of the ER, contained RNA and were thus responsible for the basophilia, and were named "ribonucleoprotein particles." Another question remained to be answered: were these particles involved in protein synthesis?

The answer was soon to come. The early work, using experiments *in vivo* and homogenates and tissue slices *in vitro*, was summarized in reviews by the three principle laboratories; that of the Zamecnik[28] group in 1949, of the Greenberg group in 1948, and that of the Borsook group in 1950. In 1952 Siekevitz, working in the Zamecnik laboratory, combined the cell fractionation technique, devised earlier by Claude, with the use of radioactive amino acids to show that radioactive alanine was incorporated predominantly into the proteins of an isolated microsomal fraction. This finding mirrored earlier work *in vivo* by Keller (1951), by the Zamecnik group, and by Borsook et al. and Hultin in 1950. Furthermore it was shown by Siekevitz, in the same paper, that an isolated microsomal fraction together with mitochondria and a supernatant fraction, resulted in the highest incorporation. This depended on oxidative metabolism, particularly on the ATP produced by the mitochondria, as found earlier with tissue slices in the three laboratories mentioned above.

A bit later, Zamecnik and Keller (1954) discovered that isolated microsomes, coupled with a soluble ATP-producing system, could generate a better and more reproducible procedure for incorporation into the microsomal proteins. Thus the system isolated *in vitro* began to duplicate the results obtained *in vivo* with tissue slices and homogenates. Finally it became obvious that the next step was to show that ribonucleoprotein particles, isolated from eukaryotes by detergent treatment, could incorporate amino acids into nascent proteins bound to the particles. And this was found almost simultaneously by Rendi and Hultin in 1959, Keller et al. in 1959, Kirsch et al. in 1960, and by Takanami in the same year. A not so inconsequential step was taken by R.B. Roberts in 1959 who suggested the "pleasantly sounding" term, ribosomes, for the particles, and of course the suggestion took hold.

THE RIBOSOME: EARLY YEARS

Let us now turn back a bit and examine ribosome structure. These structures, then still called "ribonucleoprotein particles," began to be studied actively in the early 1940s by biophysical procedures as if they were high molecular weight complexes. Thus, Taylor's group in 1942 and 1943 isolated these complexes from chick embryos and from extracts of human and rabbit brain as 71S–79S RNA-containing particles. They also noted, at that early date, that these particles could be broken up into 40S and 60S units by increasing salt and pH. Electron microscopic images, vague at that time in 1943, by Kahler and Bryand, indicated particles of about 18 μM in diameter. However, it took about 10 years for advances in the isolation and conditions for the stability of the particles, and for advances in analytical ultracentrifugation, before a more definitive picture began to emerge. The work, done mostly by Chao and Schachman in 1956 and by Petermann in 1955 and 1957, found that the particles from bacteria, yeast, and liver were quite similar, having half protein and half RNA, and with Svedberg constants of 50S and 30S in bacteria, of 60S and 80S in yeast, and of 75 to 80S in eukaryotes, with all the particles being stabilized by Mg^{2+}. Thus it became apparent that these isolated particles were the counterpart *in vitro* of the 10–15 μM diameter particles then seen by better EM methods.

However, confusion abounded due to the uncertainties rendered by the profusion of sedimentation values (20, 30, 40, 60, 80S) encountered. In time it began to dawn on researchers that the sedimentation characteristics, as registered in the ultracentrifuge, were dependent on particle concentration, on Mg^{2+}, on pH, and on salt concentration. All of these variables could lead to the unfolding of the particles and to their dimerization. By the time that the Petermann classic monograph appeared in 1964, there was sufficient knowledge to conclude that the 80S mammalian ribosomes were composed of 40S and 60S subunits, and that the 70S bacterial ribosomes were composed of 30S and 50S subunits, with all the subunits having about 50% RNA. The reason for the existence of the subunits was a mystery at that time and had to await further work, then ongoing, on the biochemistry of protein synthesis.

THE BIOCHEMISTRY

Let us now turn to work in the early 1950s, when all that was known was that radioactive amino acids ended up in the proteins of the microsome fraction, both *in vivo* and *in vitro*, and that the radioactive proteins were bound to the ribosomes, either in a "free" state as in bacteria, or as "bound" to membranes as in eukaryotes. But the biochemical mechanisms were unknown. A breakthrough came in 1956 when Hoagland, working close to the Lipmann laboratory, became intrigued by the possibility that amino acids could be "activated" in the same manner that the Lipmann group found for fatty acids, via ATP energy. He and his co-workers in the Zamecnik group found that to be the case, for the entering step for amino acid incorporation into protein was, via ATP activation, giving an aminoacyl–adenylate, catalyzed by synthetases, one for each amino acid. A suggestion of this, but not the mechanism, was made by Siekevitz in 1952. What next? It was known at this time that there was a form of RNA, not ribosomal, called soluble or sRNA, comprising 10% of total cellular RNA, which was in the high-speed supernatant after the ribosomes were sedimented. Its function was discovered soon thereafter, in 1958, by Hoagland et al. The aminoacyl–adenylate was transferred to this sRNA, which began to be called transfer or tRNA, by an esterification of the amino acid moiety, by the same aminoacyl synthetase, as was shown when the synthetase was purified by two other laboratories. Thus this same enzyme forms the aminoacyl–AMP anhydride as well as the esterification of the amino acid to tRNA. Furthermore it had become known that all purified tRNAs have a common cytidylic–cytidylic–adenylic (CCA) terminus, and it was then discovered that the amino acid was transferred to the adenylic acid moiety at the end, releasing AMP. The Hoagland group did find in the cell-free system that the radioactive amino acid on the tRNA was transferred to a microsome fraction, undoubtedly to the ribosome part of that fraction; thus the intermediacy of the amino acid–tRNA in protein synthesis was established.

It is instructive to note that in the same year (1958), Crick[5] postulated the existence of an "adaptor" molecule which would direct the activated amino acid to its proper place for protein synthesis. He thought that a trinucleotide would be sufficient, but the tRNAs contain about 75 nucleotides, and soon thereafter it was found that the tRNA *in toto* was necessary to position the amino acid. A question then arose about specificity; does it reside in the amino acid attached to tRNA, or does it reside in the tRNA molecule, in its nucleotide sequence? The answer was the latter. Chapeville et al., from Lipmann's laboratory, did an elegant experiment in 1962. They isolated cysteine–tRNA, then converted the cysteine to alanine through reduction by Raney nickel, thus obtaining an alanyl–tRNA. Fortunately at that time results came from the Nirenberg[15] and Ochoa[11] laboratories, working with synthetic polynucleotides, that a particular one, polyuridylic–guanylic (polyUG), was specific for cysteine and not for alanine incorporation into protein (more about this below). A reticulocyte ribosomal system, which had just been shown by others to incorporate amino acids into hemoglobin, was used together with a supernatant fraction and the charged tRNA, both from *E. coli*. When alanyl–tRNA plus polyUG were added, alanine was incorporated instead of cysteine. This result clearly showed that the specificity did not reside in the amino acid, for if it did, alanine would not have been incorporated into hemoglobin. Instead it resided in the tRNA, even though the "wrong" amino acid was attached. Furthermore, the use of charged tRNA from *E. coli* with a mammalian protein synthesizing system showed that the code for amino acid tRNAs was universal.

About this time, in the late 1950s and early 1960s, many questions arose and begged to be answered. Different aspects of protein synthesizing machinery were investigated, and the results were beginning to come together to provide a somewhat coherent picture of what was going on. First, there was still a question as to whether there were small peptide intermediates in the process of synthesis; this was answered conclusively by Loftfield and his colleagues in 1956, and the answer was no. Second, there was the question as to whether synthesis started from the amino or carboxyl end of the protein; again the results from Bishop et al. in Schweet's group in 1960 and from Dintzis in 1961 were definitive, since time-studies of the incorporation of amino acids into hemoglobin found that the amino-terminus was the starting point. Third, a question was raised as to whether there existed another RNA molecule besides tRNA and ribosomal RNA, since many researchers thought there should be a linkage between the DNA "code" and the eventual production of proteins which were thought to be coded by nucleotides in the DNA. The first evidence for this came in 1956 when Volkin and Astrachan found that upon phage infection of bacterial cells a metabolically unstable RNA molecule was formed, distinct from the more stable ribosomal RNA. The explanation for that finding was not apparent at that time, but the results were confirmed by various groups in 1960–1961 and the explanations were unequivocal that this newly found species of RNA, named mRNA (for messenger), was necessary for the synthesis of proteins. This idea of a messenger derived from the belief that the probable "coding" sequence in DNA was "transcribed" into an RNA molecule for the sequentially necessary "translation" step in protein synthesis.

The first evidence for a code came years later, in 1961, when Nirenberg caused quite a stir with his report at the International Congress of Biochemistry held in Moscow. He and Matthei used a bacterial enzyme, found several years earlier by Grunberg-Manago and Ochoa in 1955, that could indiscriminately synthesize polyribonucleotides—indiscriminately because the product depended on the nature of the substrates, and thus various kinds and combinations of nucleotides could be produced and purified. Using polyuridylic acid formed by the Ochoa enzyme, Nirenberg and Matthei in 1961 showed that with a bacterial ribosome system, the addition of this "RNA" resulted predominately in the synthesis of polyphenylalanine. In the same year, Lengyel in Ochoa's group found that the addition of polyadenine resulted in the formation of polylysine. You can imagine that the rush was on; many laboratories began to use many such "mRNAs", and thus the DNA code came into being (see Chapter 2 for details of the resolution of the coding problem).

Incidentally, as an aside, Ochoa received a Nobel prize for the synthesis of RNA, though later it was found that this enzyme was not the one that "transcribed" DNA into RNA. This circumstance indicates how wrong the prize can be when given for a specific piece of work. The same calamity befell Kornberg[9] for the discovery of the enzyme synthesizing DNA, for it was later learned that this enzyme was a DNA "repair" enzyme. How much better to have given the prizes to Ochoa and Kornberg for a greatly outstanding body of biochemical work; but how can one change the testament of Nobel?

Quickly it was discovered, in 1961, simultaneously by Brenner et al. and by Gros et al. that the rapidly turning-over mRNA became bound to ribosomes. Indeed, a little later, it was found by Okamoto and Takanami that it became bound to the small ribosomal subunit, while the binding of tRNA and of the newly synthesized protein occurred on the large ribosomal subunit, as discovered by Gilbert[22] in 1963. The roles for the subunits for these processes were the same for the bacterial and eukaryotic ribosomes. Also, at this time, Warner et al., in Rich's laboratory, found that the ribosomes with the nascent radioactive polypeptide chains sedimented at a heavier position in sucrose-density gradients than did the monomeric ribosomes; the term used at first was "heavy" ribosomes, which was later changed to "polysomes" by Rich. The visual nature of these sedimented polysomes was provided by the electron microscope pictures of Slayter et al., in that ribosomes appeared to be strung together on a string. The presumption was (a true presumption as it later turned out) that the string was the mRNA molecule. That this image was not an artifact of sedimentation had already been indicated many years earlier in 1956 by electron microscopic images of liver and pancreatic tissue by Palade and Siekevitz, though the reason for the pictures of ribosomes as whorls was not apparent at that early time. But this is the nature of science; a valid observation may have to wait many years before an adequate explanation is forthcoming.

The stage was now set for an intimate investigation of the process of protein synthesis; namely, was anything else required besides ribosomes, the enzyme involved in the activation of the amino acids and in the transfer of the charged amino acid to a tRNA molecule, the need for an mRNA, and the enzyme(s) involved in the actual peptide bond synthesis? The early experiments *in vitro* showed that even when aminoacyl–tRNA was included, and later, when mRNA was also included, there was still a need for the supernatant fraction for incorporation to occur. Thus, soluble factors were necessary, and the first to be discovered, in 1956, by Keller and Zamecnik, was GTP. Its addition was necessary for the step after the charging of the tRNA, and indeed an enzyme was found on the ribosomes which was a GTPase, though its involvement in protein synthesis was to come later.

In the later 1960s, many laboratories began to examine the role(s) of the soluble factors, and for convenience sake they divided the process of protein synthesis into three categories: initiation, elongation of the nascent chain, and termination. At about this time it was discovered that many bacterial proteins had methionine, alanine, or serine at the N-terminal end, with methionine being by far the most abundant. The reason for this was soon forthcoming, when Marcker and Sanger,[12] in 1964, found an N-blocked methionyl–tRNA, namely formylmet (fmet)–tRNA, and an enzyme existed which clipped off the formyl group. In the 1970s many laboratories began to dissect this initiation process, mainly because of the existence of purified components, such as the large and small ribosomal subunits, and purified soluble factors. Thus the first step found was the binding of a soluble factor, called IF-3 in bacteria, and EIF-3 in eukaryotes, to the small ribosomal subunit. This complex then bound to the complex of GTP/met-tRNA; the reason seemed to be that the system was not yet ready for the binding of the larger subunit to the smaller one. Met–tRNA is involved in this step in both the prokaryotic and eukaryotic systems, the formyl moiety being chipped off in bacteria. It appeared that in bacteria, fmet–tRNA was necessary for the recognition of the initiation mRNA codon, AUG, while in eukaryotes, met–tRNA was sufficient for this recognition.

The finding of non-codon directed binding of met–tRNA to the small subunit, even though there was no mRNA there, and hence no initiation codon, seemed puzzling. But it was certain that the small subunit bound met–tRNA in the absence of mRNA, and would not bind mRNA in the absence of bound met–tRNA. Another purified factor, called IF2, was found necessary to strengthen the complex which became small subunit—IF3–IF2–GTP–met–tRNA. This latter complex is now competent to bind mRNA, but only when another purified factor, IF4, together with ATP, is added. When this occurs, GTP is split to GDP and Pi and the large subunit becomes bound, finally giving the monomeric ribosomal complex (70S in the case of bacterial and 80S in the case of eukaryotic ribosomes) containing bound met–tRNA and mRNA. At this point, all the initiation factors are released, and it is this step which requires the hydrolysis of GTP; if nonhydrolyzable analogues of GTP are used, the large subunit cannot bind to form the ribosomal complex. So, after many years, the role of GTP was finally elucidated.

The question was then raised as to the identity of the recognition site on the ribosome for the initiator codon on mRNA. In bacterial ribosomes, Shine and Dalgarno (1974) found by sequence analysis that the small ribosomal RNA (16S in the case of bacteria) forms base pairs with a site on the mRNA adjacent to the initiator codon, and it was found soon thereafter by Steitz and Jakes (1975) that labeled mRNA from a phage was hydrogen-bonded to a similar fragment of rRNA. Another question raised was the use of fmet–tRNA and met–tRNA as signals. Why many, but not all, bacterial proteins had N-terminal methionine, but only a few eukaryotic proteins had this amino acid at the N-terminal end? In the case of bacteria, the formyl group was split off, and in some cases, various peptide lengths were also eliminated. In eukaryotic cells, where most proteins had to seek their destinations among the various organelles and membranes which are not present in bacteria, the internal architecture of the cell made it necessary, as will be described below, to clip off various lengths at the N-terminal ends, so leaving a great variety of N-terminal amino acids.

Turning to the elongation step in protein synthesis, this was worked out mainly by Lipmann's group (see the review by Lucas-Lenard and Lipmann, 1971, for a summary). They broke down the process as follows: the incoming aminoacyl–tRNA binds to the ribosome at an acceptor (A) site; the fmet–tRNA or the met–tRNA, or the peptidyl chain binds at the peptidyl (P) site. The latter links its growing peptide chain to the amino group of the aminoacyl–tRNA on the A site. It was quickly found that soluble factors, named Tu and Ts (T for transfer), as well as a G factor (for GTP) were necessary for this step. Tu and Ts, together with GTP, form a Tu–GTP complex, releasing Ts, and then, with the aminoacyl–tRNA, form an aminoacyl–tRNA–Tu–GTP complex. This complex is brought to the ribosomal A site, where, together with the mRNA bound to the small subunit, a 70S ribosome is formed. Peptide formation takes place between the A and P sites, GTP is hydrolyzed, and Tu and GDP are released from the complex, probably by Ts, thus recycling Tu for another cycle of peptide bond formation.

According to Lipmann (1967), the growing peptide chain that formed on the A site has to move over to the P site, leaving the A site vacant for the next oncoming aminoacyl–tRNA. However, this explanation has been questioned, mainly by Chetverin and Spirin in 1982, who, based on experiments by the Pestko and Spirin groups, found that only ribosomes plus mRNA plus aminoacyl–tRNA were necessary for translation to occur, though at a reduced rate, even in the absence of elongation factors and GTP. Spirin offered another explanation: that during the elongation cycle the binding of elongation factors and GTP "freezes" the complex at the translation site, and only upon GTP hydrolysis, removing the elongation factors, can synthesis proceed. In other words, the energy of hydrolysis is not necessary to move the growing polypeptide chain from one ribosomal site to another, but this could happen spontaneously by protein structural alterations. However, hydrolysis is necessary to release the elongation factors from the complex, which may have the incidental virtue of moving the polypeptide chain.

We now come to the peptide bond-forming enzyme, called peptidyltransferase. To this day I believe, this is an unknown being; it has never been purified and therefore is not characterized. But due to the work of Maden et al. in 1966, it was thought the enzyme was an integral part of the large ribosomal subunit, since neither supernatant factors nor GTP were found to be directly involved in peptide bond formation. Indeed, of all things, 30% methanol was found to couple ribosomes plus fmet–tRNA and aminoacyl–tRNA to effect a peptidyl transfer reaction between these two charged tRNAs; seemingly the alcohol can promote interaction between substrates and the catalytic center of the enzyme on the ribosome. Indeed, Hampl et al. in 1981 concluded that five proteins of the large ribosomal subunit, plus large subunit ribosomal RNA, constituted the peptidyl transferase center. The complete explanation is not yet in, but it would appear that several ribosomal proteins together bind to certain portions of the ribosomal RNA to so change the structure as to form the catalytic center. This feeling, that ribosomal RNA may be part of a catalytic site, brings to mind the speculation in 1968 of Crick and Orgel that the primitive ribosome might have been made up entirely of RNA.

One would think that the termination of protein synthesis is much less important than either initiation or elongation, but this is not so, considering the many factors which have been found to be involved, not only in termination, but in correct termination. Most of the work has been done with bacterial systems. In brief, there exist three specific "termination" codons in mRNA—UAA, UAG, and UGA—which direct the binding of protein release factors, called RF1 and RF2, to ribosomes; RF1 binds in response to UAA and UAG, while RF2 binds in response to UAA and UGA. In eukaryotes, there seems to be only one release factor, which responds to all three termination codons. Thus, this process of termination differs from chain initiation or elongation, in that the codon recognition molecule is a protein, as firmly established by Capecchi and Klein in 1969, and not tRNA. Another anomaly is the role of the enzyme, peptidyl transferase, which in termination causes the hydrolysis of peptidyl–tRNA rather than the formation of the peptide bond. Again there is a need for GTP, which interacts with another release factor, RF3, to stimulate RF1 and RF2 binding to ribosomes. In eukaryotes, where the one release factor, RF, also recognizes GTP, it was found that the binding of RF to ribosomes is stimulated by GTP or by a nonhydrolyzable analogue, but not by GDP. RF seems to promote a GTPase activity. The process is visualized as RF and GTP interacting with the ribosome when peptidyl–tRNA is in the P site, and when the termination codon is reached during the course of elongation. RF, with the aid of GTP, binds to the peptidyl transferase upon recognition of the termination codon. Then, upon GTP hydrolysis, the release factors and the completed protein are freed from the ribosome.

One may well wonder how it is that the same enzyme, peptidyl transferase, can catalyze peptide bond formation and also cause hydrolysis of the peptidyl–tRNA. Indeed, there is no doubt that *in vitro* the ribosomal enzyme can hydrolyze peptidyl–tRNA, as found by Scolnick et al. in 1970. One is reminded of the earlier work in the 1940s by Bergmann and Fruton on the proteolytic enzymes, where under certain conditions, perhaps similar to those in the ribosomes, a peptide bond is formed instead of being split. If only we could delve further would there be anything new to discover? In brief, it would appear that the release factor is the determinant, interacting with the enzyme to stimulate hydrolysis. Indeed, it has been found, using methods as diverse as antibodies against specific ribosomal proteins, ribosome depletion of specific proteins, and cross-linking of RF to ribosomal proteins, that RF interacts with some half-dozen proteins, both on the large and small subunits. Some of these proteins are those which also interact with Tu. Two of the proteins are part of the peptidyl transferase center, and the binding domain seems to be in an interface between the two ribosomal subunits.

Thus by 1980, a great deal of the overall picture of protein synthesis had emerged—its initiation, elongation, and termination—so that short summaries could be written, by Hunt, by Clark, and by Caskey (Figure 1). It is interesting that by and large the same mechanisms seemed to be operating in eukaryotes as in prokaryotes, indicating that once the process was perfected in bacteria, it was carried over into the eukaryotic realm and not much changed during the evolutionary span. Later work has involved refinements on what was learned at that time, and though some errors were found, inevitable in scientific research, the grand scheme has held up quite well.

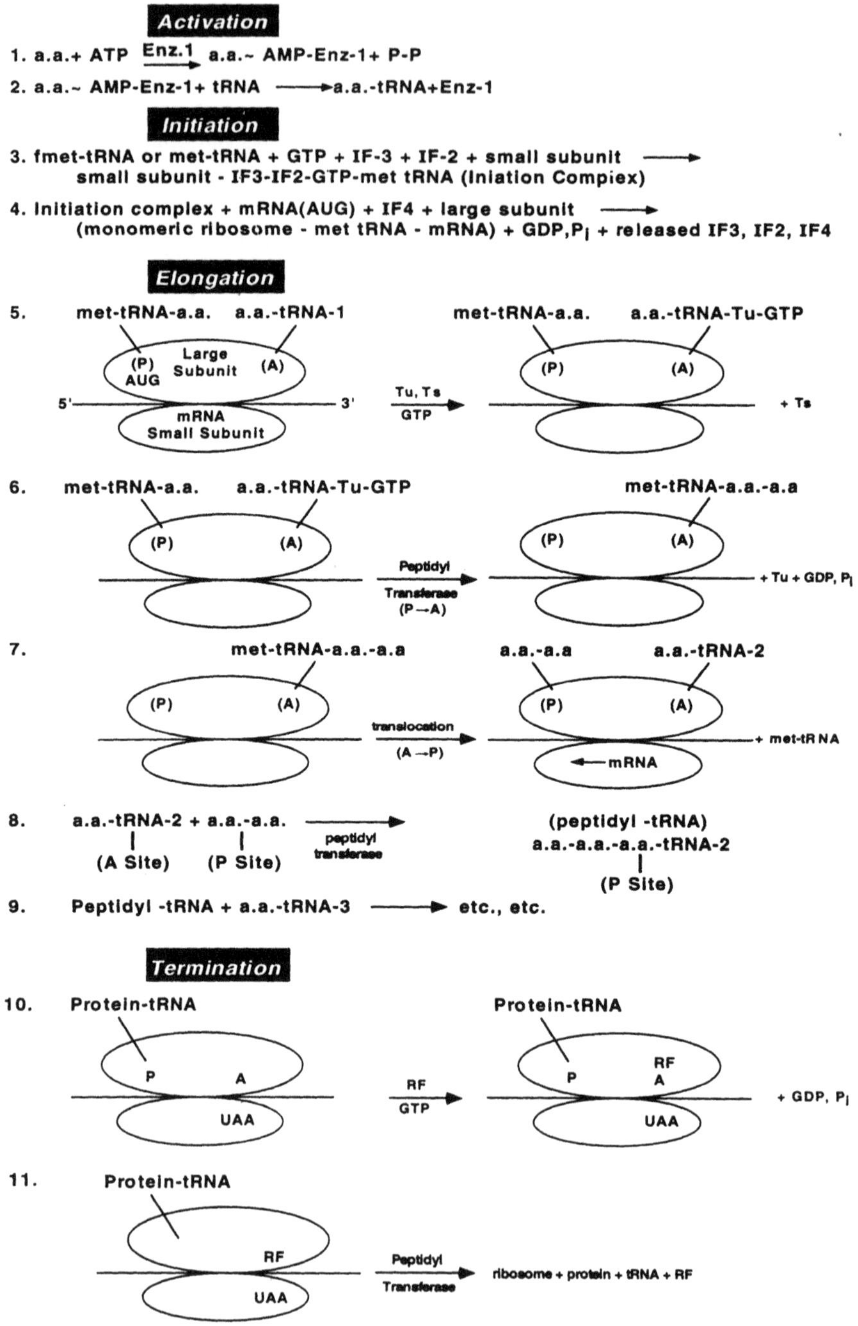

Figure 1. Steps in protein synthesis. Summary of stages described in the text.

THE RIBOSOME: STRUCTURE

I have written so far about the cell biology of protein synthesis and about the biochemistry involved in the separate steps of synthesis. The vital element in both these aspects was the identification of the ribosome as the key cellular structure. Thus, at about the same time that the biochemical experiments were ongoing, many investigators began to examine the structure of the ribosome. As mentioned, it was earlier shown in many laboratories that all ribosomes are composed of large and small subunits. I have written above on the functions of these two subunits during the process of protein synthesis. But why two subunits? There is the initiation model for the small subunit, and remarkably it was found by Kaempfer et al. in 1968, using ribosomes labeled with heavy isotopes, that during growth of *E. coli* the ribosome undergoes subunit exchange; that is, the 30S ribosomal initiation complex can detach from the 50S subunit, then become attached to other 50S ribosomal subunits. However, free 70S ribosomes do exist in bacterial cells and free 80S ribosomes do exist in mammalian cells. The question arises as to why these ribosomes should not be able to initiate protein synthesis without breaking up into their subunits. Thus it is still not entirely clear why the two-ribosomal subunit structure is necessary.

Much more work had to be done therefore to elucidate the structure–function relationships of ribosomes. The initial experiments were on the nature of the ribosomal proteins and RNA. An unexpected discovery was made by Waller and Harris in 1961 that bacterial ribosomes contain many different proteins; 21 in the 50S particle and 13 proteins in the 30S particle, all with different amino acid compositions. Conversely, also in 1961, Littauer's laboratory found, using the newly devised phenol extraction method for RNA, that ribosomal RNA was composed predominantly of only two species. A great deal of confusion abounded, due undoubtedly to RNAase action and from effects of salt and pH. But by judicious separation of the subunits in a few laboratories, it was finally acknowledged by all that the smaller bacterial subunit contains 16S RNA, the larger one having 23S RNA, while in eukaryotes the sizes were 18S and 28S, respectively.

Various methods began to be used to further examine ribosomal structure: X-ray diffraction indicated a helical structure for the RNA; electrophoretic mobilities and varied Mg^{2+} concentrations indicated that some, if not most, of the RNA is on the surface of the particle; and the use of denaturants and such salts as LiCl led to the conclusion that salt linkages and Mg^2 complexing were involved in the protein–RNA interaction. In 1963, Rosset et al. did work on a small molecular weight RNA which was also found in ribosomal phenol extracts, called it 5S RNA, and showed it was not a breakdown product of tRNA since it contained no pseudouridine nor methylated bases then known to be constituents of tRNA. Only one molecule of 5S RNA is bound, in this case to the large subunit, as compared to two tRNA molecules which are also bound.

The discovery of the multiple nature of the ribosomal proteins gave rise in the 1960s to frenzied work on the nature and possible function of these proteins. An important step was taken by Meselson and his colleagues in 1964 who were the first to use CsCl in an endeavor to break up bacterial ribosomes. They found they could split off and solubilize some proteins, leaving "core" particles unable to function in protein synthesis. Many other investigators seized upon this method, so that within a few years it was found that some of these electrophoretically purified soluble proteins were involved in the bonding of tRNA and mRNA to the "core" particles, and also involved in some of the specific steps of protein synthesis, as indicated by Traub et al. in 1966. Indeed in several laboratories, such as those of Hosokawa et al. and Kurland in 1966, and of Raskas and Staehelin in 1967, it was found possible to add back these split proteins to the "core" particles to reconstitute the 30S and 50S subunits that could then combine to form the 70S ribosome, which was active in protein synthesis—a truly remarkable feat. A big step forward was made by Lerman, Spirin, et al. in 1965 who, using graded concentrations of CsCl to dissociate the ribosome stepwise, showed that the resultant particles could be reconstituted by the addition of the proteins split off at that particular stage. The next phase in this endeavor was reached in the 1970s when it became possible to purify the individual ribosomal proteins and to characterize them for their roles in ribosomal structure and function.

The stage had been set for this advance through the introduction of new techniques, as in most scientific research. These were the purification of all the *E. coli* ribosomal proteins using new degrading reagents, new cross-linking reagents for proteins, neutron scattering, and fluorescent spectroscopy (see Brindcombe et al., 1978). Many laboratories were involved in the overall quest, with the impetus for all this being the breakthrough papers by Traub et al. in 1967 and by Nomura et al. in 1969. A significant step in the reconstitution experiments was the omission of individual proteins which were then added back. The resultant particles, either large or small, were assayed for their ability to complex with one another and to finally form the 70S particle capable of protein synthesis. The laboratories of Kurland, Wittman, Traub and Nomura, and Osawa and Spirin were working on the problems of which proteins had to be initially bound to the core particles before other proteins could be attached, which complexed to rRNA, which to mRNA, which tightened the connections between tRNA and mRNA, and which were involved in the individual steps in protein synthesis. All in all this was a truly remarkable body of work in that assembly maps could be made for each of the ribosomal subunits, and subunit interactions could be visualized during protein synthesis (Bretscher, 1968; Mizushima and Nomura, 1970; reviews by Spirin, 1969, and by Kurland, 1977).

The visualization on the surface of the ribosomal subunits of proteins whose functions were then known was shown quite elegantly by Schendorf et al. in 1974 and by Lake in 1976 by electron microscopy using antibodies to individual proteins. The beautiful pictures produced by these two groups were in general agreement, but there were some differences. However, when a comparison was made of the near-neighbors of proteins obtained by immunoelectron microscopy with those obtained by the protein cross-linking method, even more correspondence was found, if account was taken of the possible elongated nature of the proteins. Thus, the concentrated efforts of a good many individuals led, within a 10-year span from the late 1960s to the late 1970s, to a remarkable picture of ribosome structure, which together with the biochemical experiments on protein synthesis, gave a coherent view of the interactions of the proteins with one another, and with the RNA molecules instrumental in protein synthesis. Gaps remained of course, but by 1980 everyone was agreed on how structure and function had come together to produce the complicated mechanism of protein synthesis. Looking back at the ribosome, who would have thought, even in the late 1960s, that all which had been worked out was even possible.

RIBOSOMES: BIOGENESIS

At about the same time that experiments were ongoing on ribosome structure, a group of investigators were examining ribosomal biogenesis, particularly that of rRNA. The initial work was done with *E. coli* ribosomes by McCarthy et al. who in 1962, using pulse-labeling with RNA precursors, concluded that ribosomes are formed in a step-wise manner. They found that the smallest stable precursor was a 14S particle. Subsequently the nascent RNA appeared in 30S and 43S particles, and finally in the 30S and 50S ribosomal subunits. The finding was somewhat complicated because these workers were finding mRNA labeling as well as rRNA labeling. The experiments were done shortly after naturally occurring mRNA had been discovered (see above). The idea of delay points in ribosome biogenesis, at which ribosomal proteins were added, was suggested by Kono and Osawa in 1964, and proved correct. The chemistry of rRNA was also worked on at this time, and the high guanine content verified, as was the presence of pseudouridine and methylated bases in many types of ribosomal RNA.

A marked step forward was taken when the eukaryotic system was examined. Even in the 1940s it was surmised by Brachet and Caspersson that the nucleolus was the site of RNA synthesis. In 1964 this supposition was elegantly confirmed by Brown and Gurdon using an anucleolar mutant in frog, and by Miller and Beatty who, in 1969, showed beautiful electron micrographs of rRNA genes in the nucleolus strung out on DNA strands like "Christmas trees" (see Chapter 2, Figure 1). A further delineation of a step-wise procedure was uncovered by Perry in 1962 who clearly showed, in labeling and density-gradient experiments, that a large 45S RNA is first formed, that it is split to an 18S and a 32 or 35S species, and that the latter is converted to the 28S RNA (Figure 2). This process was shown to be quite complicated, as found by many laboratories; for example, during the splitting of the 45S RNA to the smaller species, many nonribosomal stretches of RNA were excised. Hybridization experiments showed that the 28S and 18S RNAs resided in the same precursor molecule and, inexplicably, that distinct and separate genes for these RNAs alternate along the DNA chain interspersed between stretches of DNA not coding for rRNA. Fascinatingly, it was discovered by hybridization experiments

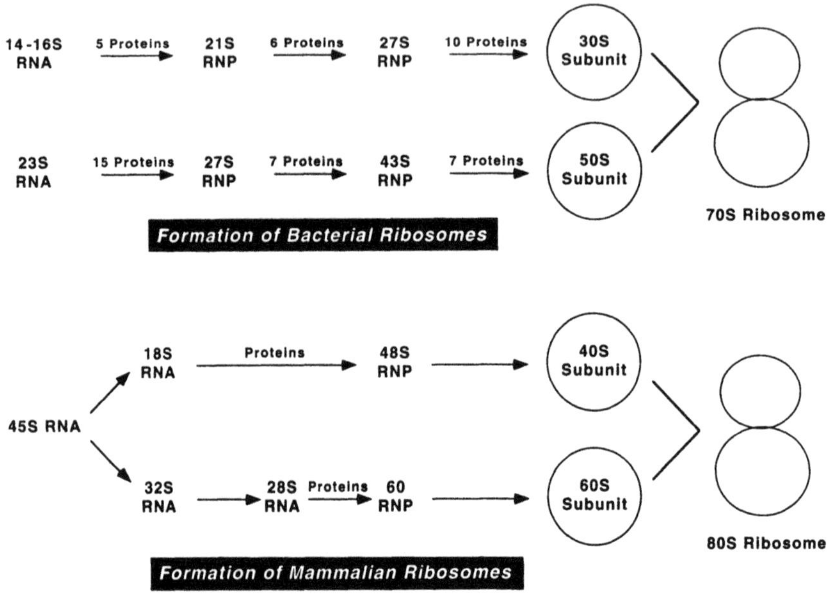

Figure 2. Abbreviated scheme of ribosome biogenesis. Details of ribosome biogenesis are given in the text.

that the genes coding for the 18S and 28S RNAs are present in clusters of thousands of copies arranged linearly along the DNA, with 18S RNA genes, a spacer region, then 28S RNA genes.

But what about the ribosomal proteins? Where are they synthesized and where do they complex with rRNAs? It seems that even the 45S precursor molecule gains some proteins in the nucleolus, and that both the 18S and 28S RNAs gain proteins in the nucleus to form the 48S and 60S particles, which are then discharged into the cytoplasm to form the 80S ribosome, as discovered by Girard et al. in 1965. As for the syntheses of these ribosomal proteins, by the late 1960s many ribosomal proteins could be identified, and it was found by several groups that they are synthesized on cytoplasmic polysomes via mRNAs secreted from the nucleus, and are then somehow transported to the nucleus and assembled there, as mentioned above.

CELL BIOLOGY: LATER YEARS

I would now like to end by going back to the beginning: the cell biology of protein synthesis. The questions to be investigated are those pertaining, not to the biochemical mechanism of protein synthesis, but to the final localization of the proteins within the cell. The answers had to come from the eukaryotic system, for here were to be found the various organelles and membrane structures. For example, it was known in the 1950s that many cells had two categories of ribosomes: membrane-bound and those called "free," both existing as polysomal structures. What was the significance of these two topologically different types of ribosomes? A good example of this is the acinar, secretory, cell of the pancreas. With this system, Siekevitz and Palade found in 1960, by labeling *in vivo* and subsequent subcellular fractionation, that chymotrypsinogen, one of the secretory proteins of the pancreas, is synthesized exclusively on ribosomes bound to the membranes of the endoplasmic reticulum. The counterpart of this system *in vitro* was the finding by Redman et al. in 1966, that another pancreatic secretory protein, amylase, is synthesized on membrane-bound ribosomes. It is then transported through the endoplasmic reticulum membrane into the cisternal space of the reticulum. Even incomplete nascent protein, truncated by the use of the protein synthesis inhibitor, puromycin, also uses the same pathway (Redman and Sabatini, 1966). These findings prompted many laboratories in the late 1960s and early 1970s to examine a large variety of proteins, both secretory and nonsecretory. By and large secreted proteins such as serum proteins made in the liver, and immunoproteins made by plasma cells, are synthesized on bound ribosomes, while a wide variety of other soluble proteins, such as ferritin, globin, and myosin are made on free ribosomes, while some membrane proteins are made on both. A good example is the work of Ganoza and Williams (1969) who found, using immunoprecipitation, that bound ribosomes from liver synthesized serum proteins, while free ribosomes from the same organ synthesized soluble liver proteins.

The next question was "How: does the protein 'know' where to go in the cell once it is synthesized?" The first indication was to come in 1972 by Milstein and co-workers who used a reticulocyte lysate system with immunoglobulin mRNA to show that a molecule could be synthesized about 3 kDa larger than the authentic light chain. When they added membranes in the form of microsomes, an N-terminal sequence was split off, and the light chain molecule, now the correct size, was bound to the microsomal membrane. Thus, they postulated that the N-terminal portion was a "signal" for the insertion of the light chain into the membrane; this was the first step in the secretory process. Soon thereafter, many papers from various groups appeared confirming and extending this hypothesis. The most extensive extension of the idea, placing it on a sure footing, was performed in the laboratory of Blobel based initially on experiments performed by Blobel and Dobberstein in 1975. The "signal hypothesis" derived from Milstein et al. was extended to state that all mRNAs for secretory proteins code for an extra N-terminal peptide, named a "signal" peptide, since it contains, in this case, the signal for penetration into the microsomal membrane. This nascent peptide inserts itself into the membrane to somehow form a tunnel in the membrane by the ribosome attachment site, dragging the rest of the nascent protein with it. The signal peptide is removed by a peptidase before the protein chain is completely synthesized, and the nascent protein then enters the cisternal (intramicrosomal) space as the first step in the secretory process. That this occurs in the pancreatic secretory process as well as in plasma cells synthesizing immunoglobulins, and that globin synthesized by and retained in reticulocytes has an mRNA having no signal codons, and thus no attachment to the membrane, are further proof of the validity of the scheme. In the next few years further validation was provided by many endocrine systems, such as the pituitary, placenta, pancreas, thyroid, and parathyroid, and by exocrine systems such as mammary gland, fibroblasts synthesizing collagen, and the hepatocyte. In all cases, evidence was present for an extra N-terminal peptide necessary for synthesis and secretion, at the correct membrane site, of the various proteins.

But what are these signal peptides? In the 1980s a great deal of work was carried out on them and their occurrence. They were found in all types of eukaryotic and prokaryotic cells having the same function. But perplexedly, when the sequences of these peptides were examined, there were no regions of strict homology among eukaryotes between the various membrane types they were instrumental in traversing, or between eukaryotes and prokaryotes. The cleavage site was the only one in which some conservation was observed. However, there were recognizable regions with specific characteristics shared by all the cleaved peptides. There was a 5–7 amino acid residue near the cleavage site with rather high polarity; another region had from 7–10 amino acids of a hydrophobic nature, and indeed hydrophobicity seemed to be the major requirement in this region; another region near the N-terminal end carried a net positive charge. Despite these defining characteristics, it is still difficult to relate them to functional roles since there seems to be no consistency with regards to protein/protein interactions. Indeed, in some cases the "signal" peptide was found to reside, not at the N-terminus, but in an interior region of the protein. A possible answer to this dilemma was the finding of a ribonucleoprotein particle consisting of a six-protein subunit, and a 7S RNA, which seemed to be involved in signal recognition and hence was named the signal recognition particle or SRP. A 72 kDa membrane protein, called a docking protein, was found to bind to the SRP, and it is this complex that plays a role in ensuring that the signal peptide begins to traverse the membrane, perhaps by destabilizing its lipids and allowing a "hole" to appear. A good review of this period (1985) is by Wickner and Lodish.

Finally, what about mitochondria and chloroplasts? Way back, in 1890, it was postulated that mitochondria are autonomous organelles capable of self-replication within the cell. We now know this is not so—that they are semi-autonomous. It was shown quite a few years ago (see Chapter 4) that DNA coding for mitochondrial proteins resides both in the mitochondria and in the nucleus; indeed, mitochondrial DNA codes for only a few mitochondrial proteins. A good example is the cytochrome oxidase complex, made up of nine polypeptides, of which six are coded for in the nucleus and synthesized by cytoplasmic ribosomes, and three are coded by mitochondrial DNA and synthesized by mitochondrial ribosomes. Yes, mitochondria, as well as chloroplasts, have DNA and ribosomes, which in the case of mitochondria resemble those of bacteria in being smaller in size than cytoplasmic ribosomes. With both these organelles, those proteins synthesized in the cytoplasm and destined for the organelles also have transit sequences similar to those mentioned above for the intracellular membranes. In addition, to confuse matters, some mitochondrial proteins, such as cytochrome *c*, are synthesized by cytoplasmic ribosomes but have no signal peptide, and hence are described as being moved into the membrane by a posttranslational mechanism. In all the mitochondrial and chloroplast cases cytoplasmic receptors have been found for the cytoplasmically synthesized proteins similar to SRP, and membrane receptors like the docking protein. As can be seen, by the middle 1980s the problem of how proteins specifically get into mitochondria and chloroplasts had not been solved.

EPILOGUE

I will end here, not because all research has been done, and all is known, but, as in the beginning, my arbitrary point of ending has been reached. The highlights, as I have seen them, have been pointed out, and, frankly, so much more could have been written to occupy a book, which I am too tired even to contemplate!

I should add that historians of science, non-scientists, have begun to delve into what might be called the personal stories in scientific research. A relevant one regarding protein synthesis is by Rheinberger, who through perusing personal papers and by interviews, has written a history of the early days of protein synthesis, mainly about the work of the Zamecnik group.

REFERENCES

Bergmann, M. & Fruton, J.S. (1941). The specificity of proteinases. Adv. Enzymol. 1, 63–98.

Bergmann, M. & Fruton, J.S. (1944). Significance of coupled reactions for the enzymatic hydrolysis and synthesis of proteins. Ann. N.Y. Acad. Sci. 45, 409–423.

Bishop, J., Leahy, S., & Schweet, R. (1960). Formation of the peptide chain of hemoglobin. Proc. Natl. Acad. Sci. USA 46, 1030–1038.

Blobel, G. & Dobberstein, B. (1975). Transfer of proteins across membranes. I & II. J. Cell Biol. 67, 835–851; 852–862.

Borsook, H. (1950). Protein turnover and incorporation of labeled amino acids into tissue proteins *in vivo* and *in vitro*. Physiol. Rev. 30, 206–219.

Borsook, H. (1956). Biosynthesis of peptides and proteins. J. Cell. Comp. Physiol. 47 Suppl., 35–80.

Borsook, H., Deasy, C.L., Haagen-Smit, A.J., Keighley, C., & Lowy, P.H. (1950). Metabolism of ^{14}C-labeled glycine, L-histidine, L-leucine and L-lysine. J. Biol. Chem. 187, 839–848.

Borsook, H. & Keighley, C. (1935). Continuing metabolism of nitrogen in mammals. Proc. Roy. Soc. London, B. 118, 488–501.

Brachet, J. (1938). Recherches sur la synthese de l'Acide thymonucleique pendant le developpement d'Oeuf d'Oursin. Archiv. de Biolog. 44, 519–576.

Brachet, J. (1942). La Localisation des acides pentosnucleiques dans les tissus animaux et les Oeufs d'Amphibiens en voie de developpement. Arch. Biolog. 53, 207–257.

Brachet, J. (1950). Localization and role of ribonucleic acid in the cell. Ann. N.Y. Acad. Sci. 50, 861–869.

Brenner, S., Jacob, F., & Meselson, M. (1961). An unstable intermediate carrying information from genes to ribosomes for protein synthesis. Nature, Lond. 190, 576–581.

Bretscher, M.S. (1968). Translocation in protein synthesis; a hybrid structural model. Nature, Lond. 218, 675–677.

Brindcombe, M.G., Stöffler, G., & Wittman, H.G. (1978). Ribosome structure. Annu. Rev. Biochem. 47, 217–249.

Brown, D.D. & Gurdon, J.M. (1964). Absence of ribosomal RNA synthesis in the anucleolate mutant of *Xenopus laevis*. Proc. Natl. Acad. Sci. USA 51, 139–146.

Capecchi, M.R. & Klein, H.A. (1969). Characterization of three proteins involved in polypeptide chain initiation. Cold Spring Harbor Sym. Quant. Biol. 34, 469–477.

Caskey, C.T. (1980). Peptide chain termination. Trends Biochem. Sci. 5, 178–181.

Caspersson, T.I. (1941). Studien über den eiweissumsatz der zelle. Naturwiss. 29, 33–43.

Caspersson, T.I. (1947). Relations between nucleic acid and protein synthesis. Soc. Exp. Biol. 1, 127–151.

Caspersson, T.I. & Schultz, J. (1939). Pentose nucleotides in the cytoplasm of growing tissues. Nature, Lond. 143, 602–603.

Chao, F.-C. & Schachman, H.K. (1956). The isolation and characterization of the macromolecular ribonucleoproteins from yeast. Arch. Biochem. Biophys. 61, 220–230.

Chapeville, F., Lipmann, F., von Ehrenstein, G., Weisblum, B., Ray, W.J. Jr., & Benzer, S. (1962). On the role of soluble ribonucleic acid in coding for amino acids. Proc. Natl. Acad. Sci. USA 48, 1086–1092.

Chetverin, A.B. & Spirin, A.S. (1982). Bioenergetics and protein synthesis. Biochem. Biophys. Acta 683, 153–179.

Clark, B. (1980). The elongation step of protein biosynthesis. Trends Biochem. Sci. 5, 207–210.

Claude, A. (1938). A fraction from normal chicken embryo similar to the tumor-producing factor of chicken tumor. Proc. Soc. Exp. Biol. Med. 89, 398–403.

Claude, A. (1941). Particulate components of cytoplasm. Cold Spring Harbor Sym. Quant. Biol. 9, 263–270.

Claude, A. (1943). The constitution of protoplasm. Science 97, 451–456.

Crick, F.H.C. (1958). On protein synthesis. Soc. Exp. Biol. Sym. London, XII, 138–163.

Crick, F.H.C. (1968). The origin of the genetic code. J. Mol. Biol. 38, 367–379.

Davidson, J.N. & Waymouth, C. (1943). Ribonucleic acids in animal tissues. Nature, Lond. 152, 47–48.

Dintzis, H.M. (1961). Assembly of peptide chains of hemoglobin. Proc. Natl. Acad. Sci. USA 47, 247–261.

Frantz, I.D. Jr., Zamecnik, P.C., Reese, J.W., & Stephenson, M.L. (1948). Effect of dinitrophenol on the incorporation of alanine labeled with radioactive carbon into the proteins of slices of normal and malignant rat liver. J. Biol. Chem. 174, 773–774.

Fruton, J.S. (1950). Role of proteolytic enzymes in biosynthesis of peptide bonds. Yale J. Biol. Med. 22, 263–271.

Ganoza, M.C. & Williams, C.A. (1969). *In vitro* synthesis of different categories of specific proteins by membrane-bound and free ribosomes. Proc. Natl. Acad. Sci. USA 63, 1370–1374.

Gilbert, W. (1963). Polypeptide synthesis in *E. coli*. II. The polypeptide chain and sRNA. J. Mol. Biol. 6, 389–403.

Girard, M., Latham, H., & Darnell, J.E. (1965). Entrance of newly formed mRNA and ribosomes into HeLa cell cytoplasm. J. Mol. Biol. 11, 187–201.

Greenberg, D.M., Friedberg, F., Schulman, M.P., & Winnick, T. (1948). Studies on the mechanism of protein synthesis with radiocarbon-labeled compounds. Cold Spring Harbor Sym. Quant. Biol. 13, 113–124.

Gros, F., Hiatt, H., Gilbert, W., Kurland, C.G., Rosebrough, R.W., & Watson, J.D. (1961). Unstable RNA revealed by pulse-labeling of *E. coli*. Nature, Lond. 190, 581–585.

Grunberg-Manago, M. & Ochoa, S. (1955). Enzymatic synthesis and breakdown of polynucleotides: Polynucleotide phosphorylase. J. Amer. Chem. Soc. 77, 3165–3166.

Hampl, H., Schulze, H., & Nierhaus, K.H. (1981). Ribosomal components from *E. coli* 50S subunits involved in the reconstitution of peptidyltransferase activity. J. Biol. Chem. 256, 2284–2288.

Hoagland, M.B., Keller, E.B., & Zamecnik, P.C. (1956). Enzymatic carboxyl activation of amino acids. J. Biol. Chem. 218, 345–358.

Hoagland, M.B., Stephenson, M.L., Scott, J.F., Hecht, L.I., & Zamecnik, P.C. (1958). A soluble ribonucleic acid intermediate in protein synthesis. J. Biol. Chem. 231, 241–257.

Hosokawa, K., Fujimura, R.K., & Nomura, M. (1966). Reconstitution of functionally active ribosomes from inactive subparticles and proteins. Proc. Natl. Acad. Sci. USA 55, 198–204.

Hultin, T. (1950). Incorporation of ^{15}N-labeled glycine into liver fractions of newly-hatched chicks. Exp. Cell Res. 1, 376–381.

Hunt, T. (1980). The initiation of protein synthesis. Trends Biochem. Sci. 5, 178–181.

Kaempfer, R., Meselson, M., & Raskas, H. (1968). Cyclic dissociation into stable subunits and re-formation of ribosomes during bacterial growth. J. Mol. Biol. 31, 277–289.

Kahler, H. & Bryand, W.R. (1943). Ultracentrifugal studies of some complexes obtained from mouse milk, mammary tumor and other tissues. J. Nat. Cancer Inst. 4, 37–45.

Kalckar, H. (1941). The nature of energetic coupling in biological syntheses. Chem. Rev. 28, 71–178.

Keller, E.B. (1951). Turnover of proteins of cell fractions of adult rat liver *in vivo*. Fed. Proc. 10, 206.

Keller, E.B. & Zamecnik, P.C. (1956). Effect of guanosine diphosphate and triphosphate on the incorporation of labeled amino acids into proteins. J. Biol. Chem. 221, 45–59.

Keller, E.B., Zamecnik, P.C., & Loftfield, R.I. (1954). Role of microsomes in the incorporation of amino acids into proteins. J. Histo. Cytochem. 2, 378–386.

Kirsch, J.F., Siekevitz, P., & Palade, G.E. (1960). Amino acid incorporation *in vitro* by ribonucleoprotein particles detached from guinea-pig liver microsomes. J. Biol. Chem. 235, 1419–1424.

Kono, M. & Osawa, S. (1964). Intermediary steps of ribosome formation in *E. coli*. Biochim. Biophys. Acta 87, 326–334.

Kurland, C.G. (1966). The requirements for specific sRNA binding by ribosomes. J. Mol. Biol. 18, 90–108.

Kurland, C.G. (1977). Structure and function of the bacterial ribosome. Annu. Rev. Biochem. 46, 173–200.

Lake, J.A. (1976). Ribosome structure determined by electron microscopy of *E. coli* small subunits, large subunits and monomeric ribosomes. J. Mol. Biol. 105, 131–159.

Lengyel, P., Speyer, J.F., & Ochoa, S. (1961). Synthetic polynucleotides and the amino acid code. Proc. Natl. Acad. Sci. USA 47, 1936–1942.

Lerman, M.L., Spirin, A.S., Gardova, L.P., & Golov, V.F. (1965). Studies on the structure of ribosomes. II. Stepwise dissociation of protein from ribosomes by caesium chloride and the re-assembly of ribosomal-like particles. J. Mol. Biol. 15, 268–280.

Lipmann, F. (1941). Metabolic generation and utilization of phosphate bond energy. Adv. Enzymol. 1, 99–162.

Lipmann, F. (1967). Peptide bond formation in protein synthesis. In: Regulation of Nucleic Acid and Protein Biosynthesis (Koningsberger, V.V. & Bosch, L., Eds.), pp. 177–186. Elsevier, Amsterdam.

Littauer, U.Z. (1961). Biochemical activity and structural aspects of ribonucleic acids. In: Protein Biosynthesis (Harris, R.J.C., Ed.), pp. 143–162. Academic Press, New York.

Loftfield, R.B., Groyer, J.W., & Stephenson, M.L. (1953). Possible role of proteolytic enzymes in protein synthesis. Nature, Lond. 171, 1024–1027.

Loftfield, R.B. & Harris, A.C. (1956). Participation of free amino acids in protein synthesis. J. Biol. Chem. 219, 151–159.

Loftfield, R.B. & Eigner, E.A. (1958). The time required for the synthesis of ferritin in rat liver. J. Biol. Chem. 231, 925–943.

Lucas-Lenard, J. & Lipmann, F. (1971). Protein biosynthesis. Annu. Rev. Biochem. 40, 410–448.

Maden, B.E.H., Traut, R.R., & Monro, R.E. (1966). Ribosome-catalyzed peptidyl transfer; the polyphenylalanine system. J. Mol. Biol. 35, 333–345.

Marcker, K. & Sanger, F. (1964). *N*-formyl methionyl-sRNA. J. Mol. Biol. 8, 835–840.

McCarthy, R.J., Britten, R.J., & Roberts, R.B. (1962). The synthesis of ribosomes in *E. coli*. III. Synthesis of ribosomal RNA. Biophys. J. 2, 57–82.

Meselson, M., Nomura, M., Brenner, S., Davern, C., & Schlessinger, D. (1964). Conservation of ribosomes during bacterial growth. J. Mol. Biol. 9, 696–711.

Miller, O.L. & Beatty, B.R. (1969). Visualization of nucleolar genes. Science 164, 955–957.

Milstein, C., Brownlee, G.G., Harrison, T.M., & Matthews, M.B. (1972). A Possible precursor of immunoglobulin light chains. Nature, Lond. 239, 117–120.

Mizushima, S. & Nomura, M. (1970). Assembly mapping of 30S ribosomal proteins from *E. coli*. Nature, Lond. 226, 1214–1218.

Nirenberg, M.W. & Matthei, J.A. (1961). The dependence of cell-free protein synthesis in *E. coli* upon naturally occurring or synthetic polyribonucleotides. Proc. Natl. Acad. Sci. USA 47, 1588–1602.

Nomura, M., Mizushima, S., Osaki, M., Traub, P., & Lowry, C.V. (1969). Structure and function of ribosomes and their molecular components. Cold Spring Harbor Sym. Quant. Biol. 34, 49–61.

Orgel, L.F. (1968). Evolution of the genetic apparatus. J. Mol. Biol. 38, 381–393.

Okamoto, T. & Takanami, M. (1963). Interaction of ribosomes and some synthetic polyribonucelotides. Biochim. Biophys. Acta 68, 325–327.

Palade, G.E. (1955). A small particulate component of the cytoplasm. J. Biophys. Biochem. Cytol. 1, 59–68.

Palade, G.E. (1956). The endoplasmic reticulum. J. Biophys. Biochem. Cytol. 2 Suppl., 85–98.

Palade, G.E. & Siekevitz, P. (1956). Liver microsomes: An integrative morphological and biochemical study. J. Biophys. Biochem. Cytol. 2, 171–198.

Palade, G.E. & Siekevitz, P. (1956). Pancreatic microsomes: An integrative morphological and biochemical study. J. Biophys. Biochem. Cytol. 2, 671–690.

Perry, R.P. (1962). The cellular sites of ribosomal 45S RNA. Proc. Natl. Acad. Sci. USA 48, 2179–2186.

Petermann, M.L. (1964). Physical and Chemical Properties of Ribosomes. Elsevier Press, Amsterdam.

Petermann, M.L. & Hamilton, M.G. (1955). A stabilizing factor for cytoplasmic nucleoproteins. J. Biophys. Biochem. Cytol. 1, 469–472.

Petermann, M.L. & Hamilton, M.G. (1957). The purification and properties of cytoplasmic ribonucleoproteins from rat liver. J. Biol. Chem. 224, 725–736.

Porter, K.R. (1954). Electron microscopy of basophilic components of cytoplasm. J. Histo. Cytochem. 2, 346–373.

Porter, K.R., Claude, A., & Fullam, E. (1945). A study of tissue culture cells by electron microscopy. J. Exp. Med. 81, 233–246.

Raskas, H.J. & Staehelin, T. (1967). Messenger and sRNA binding by ribosomal subunits and reconstituted ribosomes. J. Mol. Biol. 23, 89–97.

Redman, C.M. & Sabatini, D.D. (1966). Vectorial discharge of peptides released by puromycin from attached ribosomes. Proc. Natl. Acad. Sci. USA 56, 608–615.

Redman, C.M., Siekevitz, P., & Palade, G.E. (1966). Synthesis and transfer of amylase in pigeon pancreatic microsomes. J. Biol. Chem. 241, 1150–1158.

Rendi, R. & Hultin, T. (1959). *In vitro* incorporation of ^{14}C amino acids into partially purified ribonucleoprotein particles. Exp. Cell Res. 18, 542–553.

Rheinberger, H.-J. (1993). Experiments and orientation: Early systems of *in vitro* protein synthesis. J. Hist. Biol. 26, 443–471.

Rosset, R., Julien, J., & Monier, R. (1963). A propos de la presence d'acide ribonucleique de faible poids moleculaire dans les ribosomes d'*E. coli*. Biochem. Biophys. Acta 68, 653–656.

Schendorf, T., Zeichhardt, H., & Stöffler, G. (1974). Determination of location of proteins L14, L17, L18, L19, L22, L23 on the surface of the 50S ribosomal subunit of *E. coli* by immune electron-microscopy. Mol. Gen. Genet. 134, 187–208.

Schoenheimer, R. (1942). Dynamic State of Body Constituents. Harvard University Press, Cambridge, MA.

Schweet, R., Lamfrom, H., & Allen, E. (1958). The synthesis of hemoglobin in a cell-free system. Proc. Natl. Acad. Sci. USA 44, 1029–1035.

Scolnick, E., Milman, G., Rosman, M., & Caskey, C.T. (1970). Transesterification by peptidyltransferase. Nature, Lond. 225, 152–154.

Shine, J. & Dalgarno, L. (1974). The 3′-terminal sequence of *E. coli* 16S ribosomal RNA: Complimentary to nonsense triplets and ribosomal binding sites. Proc. Natl. Acad. Sci. USA 71, 1342–1346.

Siekevitz, P. (1952). Uptake of radioactive alanine *in vitro* into the proteins of rat liver fractions. J. Biol. Chem. 195, 549–565.

Siekevitz, P. & Palade, G.E. (1960). A cytochemical study on the pancreas of the guinea-pig. V. *In vivo* incorporation of leucine-1-^{14}C into the chymotrypsinogen of various cell fractions. J. Biophys. Biochem. Cytol. 7, 619–630.

Slayter, H.S., Warner, J.R., Rich, A., & Hall, C.E. (1963). The visualization of ribosomal structure. J. Mol. Biol. 7, 652–657.

Spirin, A.S. (1969). A model of the functioning ribosome: Locking and unlocking of the ribosomal subparticles. Cold Spring Harbor. Sym. Quant. Biol. 34, 197–207.

Steitz, J.A. & Jakes, K. (1975). How ribosomes select initiator regions of mRNA: Base pair formation between the 3′-terminus of 16S RNA during initiation of protein synthesis in *E. coli*. Proc. Natl. Acad. Sci. USA 72, 4734–4738.

Takanami, M. (1960). A stable ribonucleoprotein for amino acid incorporation. Biochem. Biophys. Acta 39, 318–326.

Taylor, A.G., Sharp, D.G., Beard, D., & Beard, J.W. (1942). Isolation and properties of a component of normal chick embryo tissue. J. Infect. Dis. 71, 115–126.

Taylor, A.G., Sharp, D.G., & Woodhall, R. (1943). Macromolecular components of normal embryonic and adult brain tissue. Science 97, 226–227.

Traub, P., Hosokawa, K., Craven, G.R., & Nomura, M. (1967). Structure and function of *E. coli* ribosomes, IV: Isolation and characterization of functionally active ribosomal proteins. Proc. Natl. Acad. Sci. USA 58, 2430–2436.

Traub, P., Nomura, M., & Tu, L. (1966). Physical and functional heterogeneity of ribosomal proteins. J. Mol. Biol. 19, 215–218.

Volkin, E. & Astrachan, L. (1956). Phosphorus incorporation in *E. coli* ribonucleic acid after infection with bacteriophage T2. Virology 2, 149–161.

Waller, J.-P. & Harris, J.I. (1961). Studies on the composition of the protein from *E. coli* ribosomes. Proc. Natl. Acad. Sci. USA 47, 18–23.

Warner, J.R., Rich, A., & Hall, C.E. (1962). Electron microscopy of ribosomal clusters synthesizing hemoglobin. Science 138, 1399–1403.

Warner, J.R., Knopf, P.M., & Rich, A. (1963). A multiple ribosomal structure in protein synthesis. Proc. Natl. Acad. Sci. USA 49, 122–129.

Wickner, W.T. & Lodish, H.F. (1985). Multiple mechanisms of protein insertion into and across membranes. Science 230, 400–407.

Zamecnik, P.C. & Frantz, I.D. Jr. (1949). Peptide bond synthesis in normal and malignant tissues. Cold Spring Harbor Sym. Quant. Biol. 14, 109–208.

Zamecnik, P.C., Frantz, I.D. Jr., Loftfield, R.I., & Stephenson, M.L. (1948). Incorporation *in vitro* of radioactive carbon from carboxyl-labeled dl-alanine and glycine into proteins of slices of normal and malignant rat liver. J. Biol. Chem. 175, 294–314.

Zamecnik, P.C. & Keller, E.B. (1954). Relations between phosphate energy donors and incorporation of labeled amino acids into proteins. J. Biol. Chem. 209, 337–354.

Chapter 6

STRUCTURAL BIOLOGY:
YESTERDAY, TODAY, AND TOMORROW

Iain D. Campbell

INTRODUCTION

The study of the structures of biological macromolecules, at atomic level resolution, is now usually referred to as "structural biology." The development of this field, and its tools, make a fascinating story—one where new discoveries in physics were rapidly applied to biological systems by intrepid optimists. A notable example is how the heroic efforts of Max Perutz[32] and colleagues (1985) led, after decades of labor, to the first protein structure; this was achieved by interpreting the patterns of spots arising when protein crystals are irradiated with X-rays. Today, new macromolecular structures are appearing at a rate of several hundred per year, and X-ray crystallography has been joined by two other methods capable of giving high resolution 3D structures–electron microscopy and nuclear magnetic resonance (NMR).

The results of structural biology are at the core of almost all modern biochemistry—they form the framework on which we hang most other measurements and ideas. The importance of the field has been recognized by the award of a large number of Nobel prizes to its practitioners. Structural knowledge has had a major impact on our understanding of all biological macromolecules and their complexes, including DNA, membranes, and oligosaccharides, but proteins have the most complex structures that are defined precisely by the DNA "blueprint." For this reason, and for brevity, I will concentrate on proteins here. Most aspects of structural biology have taken place within the last 50 years, including the "breakthrough," "consolidation," and "exploitation" stages favored by military strategists (Phillips, 1971). I thus extend the period discussed to somewhat beyond the 1960–1990 format of this book. This article indicates how structural biology developed (yesterday), what the current situation is (today), and possible future developments (tomorrow).

THE DEVELOPMENT OF STRUCTURAL BIOLOGY (YESTERDAY)

Early microscopes made by Hooke and van Leeuwenhoek in the seventeenth century represent the first attempts to explore the properties of biological materials beyond the limits of the naked eye (see e.g. Bracegirdle, 1989). Light, however, has the fundamental problem of having a wavelength that is several orders of magnitude too long to observe macromolecules in detail.

Observation of macromolecular structure thus depended on new developments in physics, nearly all of which are relatively recent. Röentgen discovered X-rays and J. J. Thomson electrons about 100 years ago. Chadwick discovered the neutron in 1932 and the phenomenon of nuclear magnetic resonance was demonstrated 50 years ago. These discoveries were all rapidly exploited and applied to biological systems, although a backward look at some of the early experiments suggests that many were initially carried out more in a spirit of optimism than realism. That said, it is likely that even these early optimists would be astonished at the current achievements of modern structural biology and the beauty of the structures revealed.

Diffraction

Experiments on ordered arrays, or crystals, of macromolecules have been particularly important in the development of structural biology. If radiation of suitable wavelength is directed at an ordered array, diffraction patterns can be observed. That crystals had an internal periodic structure was first surmised by seventeenth century scientists like Kepler and Hooke, but this remained a hypothesis until it was tested with X-rays, at the suggestion of von Laue in 1912. This led quickly to the first structure determination by X-ray methods, that of sodium chloride by W. L. Bragg in 1913.

Urease was the first protein to be crystallized by Sumner in 1928, but this was rapidly followed by the crystallization, by Northrop, of pepsin, trypsin, and chymotrypsin. In his book, *Protein Structure* (1992), Max Perutz describes how, in

1934, J. D. Bernal and Dorothy Crowfoot (later Hodgkin[30]) placed a crystal of pepsin in an X-ray beam to see if it gave a diffraction pattern.

> It was an unpromising experiment because it had already been proved that protein crystals gave no diffraction pattern. This was only to be expected because the great German chemist Richard Willstätter and his pupils had shown that proteins are colloids of random structure, and that the enzymatic activity of J. H. Northrop's crystalline pepsin did not reside in the protein.... Besides, even if the German chemists were wrong, and a diffraction pattern were obtained, it would clearly be impossible to deduce from it, structures of molecules as large and complex as proteins.
>
> Contrary to reason or perhaps because they had not read the literature Bernal and Crowfoot discovered that pepsin crystals *did* give a diffraction pattern. It was made up of sharp reflections that extended to spacings of the order of interatomic distances, showing that pepsin was not a colloid of random coils, but an ordered three-dimensional structure in which most of its 5,000 atoms occupy definite places. Their observation opened up the subject of protein crystallography.

Perutz himself joined Bernal's laboratory in 1936. David Eisenberg (1994), in an invited tribute to Max Perutz on the occasion of his 80th birthday, tells how Max was still without a thesis project after a year in Cambridge. He returned to Austria for a summer vacation, where Haurowitz, a distant relative, suggested hemoglobin since he had noted that oxygen changed the crystalline form of horse deoxyhaemoglobin. Perutz returned to Cambridge with the goal of understanding these changes. He did not realize that it would take 22 years even to make a start. The problem was that there was then no way to interpret the observed diffraction patterns; computers or even calculators were not around to help!

Numerous excellent accounts are now available on the early development of X-ray diffraction techniques (Perutz, 1985; Rossman, 1994) and I do not propose to repeat them in detail now. Many obstacles had to be overcome, not least the difficulty of computing the "phase" of the information in the diffraction pattern. This information is lost when the diffraction data are collected on a photographic film. Suffice to say, here, that Perutz and colleagues elegantly solved the "phase problem" by use of the "multiple isomorphous replacement" method where several heavy atoms, like mercury, are introduced into protein crystals. Collection of the diffraction data, with and without such metals, can be interpreted to yield phase information, thus allowing "electron density" representations of the protein molecule to be calculated. This breakthrough led to the first low resolution structure of myoglobin by Kendrew[33] et al. (1958) (see Figure 1). Their paper makes interesting reading today and has impressive style:

> the model must be imagined as clothed in an invisible integument of side chains so thick that neighbouring chains in reality touch.

In spite of this, the result was, at first, a disappointment (Perutz, 1964):

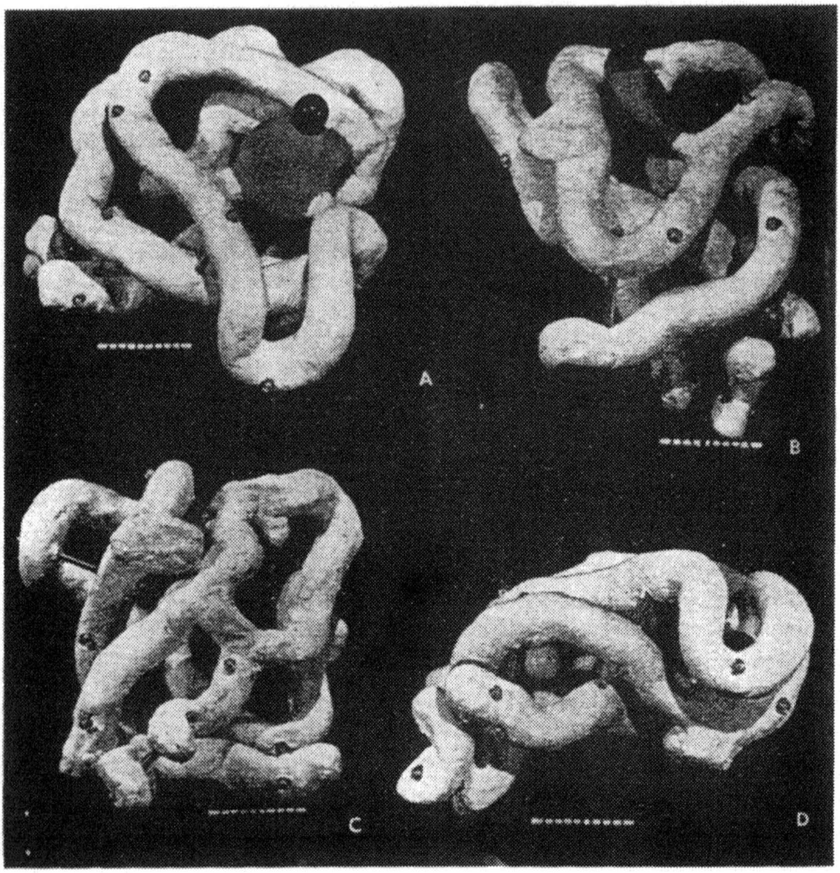

Figure 1. Views of a model of sperm whale myoglobin, built to fit low resolution electron density maps, reconstructed from X-ray diffraction patterns. The grey disc is the heme group. The white dashed lines indicate the spatial resolution, with each dash representing 1 Å. The "phase problem" was overcome by incorporating various heavy atoms into the crystal, including mercury and gold (from Kendrew et al., 1958, with permission).

> Could the search for ultimate truth really have revealed so hideous and visceral-looking an object? Was the nugget of gold a lump of lead? Fortunately like many other things in nature myoglobin gains in beauty the closer you look at it.

There were, of course, essential parallel developments going on in biochemistry in the decades that the crystallographers were wrestling with their data. Not least was new knowledge about the nature of proteins. By the late 1950s, much of protein chemistry had been worked out and, through the pioneering work of Sanger,[12] the amino acid sequences of several proteins were known. This meant that specific amino acid chains could be built into the better and higher resolution electron density maps that were later produced. It is, however, worth noting that the magnitude of the task increases dramatically as one strives for higher resolution. Kendrew et al. (1958) computed the structure shown in Figure 1 from about 400 reflections (diffraction spots on the film). At this resolution, only the overall shape of the protein can be observed. To achieve 1.5 Å resolution, when individual atoms can be resolved, requires analysis of about 20,000 reflections for a protein the size of myoglobin.

Even the X-ray crystallographers were taken by surprise at the rapid pace at which protein structures were then solved after the myoglobin breakthrough (see e.g. Bragg, 1967). By 1967, not only myoglobin but also lysozyme, ribonuclease, chymotrypsin, and carboxypeptidase were solved to a resolution of 2.0 Å or better (Stryer, 1968). Knowledge of these structures meant that other chemical and physicochemical experiments on enzymes could be interpreted in new and unprecedented detail.

X-rays have been, and will remain, the dominant radiation source for protein crystallography, but the use of *neutrons* is worthy of brief mention. Thermal neutrons from a reactor have an energy that corresponds to a wavelength of around 1 Å, about the same as many X-ray sources. However, while X-rays are scattered by electrons, neutrons are scattered by nuclei. The main advantage of using neutrons over X-rays is that H-atoms can be observed readily; moreover, the hydrogen isotopes, ^{1}H and ^{2}H, can be distinguished. The feasibility of using neutrons in protein crystallography was first demonstrated by Schoenborn (1969) and the technique has been very useful; for example, to observe hydrogen positions in mechanistically important enzyme residues (Kossiakof, 1985).

Electron Microscopy

The electron microscope was developed mainly by Ernst Ruska in Berlin in the 1930s. An interesting historical account is given in his Nobel lecture (Ruska, 1987). The following statement suggests that the lot of the scientist was at least as hard then as now:

> After obtaining my degree (early 1931), the economic situation had become very difficult in Germany and it seemed impossible to find a satisfactory position at a University or industry. Therefore I was glad to continue my unpaid position as doctorand in the high voltage institute.

Ruska demonstrated that a previous theoretical idea by Busch (1927) on electron trajectories could be put into practice. He constructed, in April 1931, an apparatus with two coils that acted as lenses on an electron beam. This equipment can be considered as the first electron microscope, although its total magnification of about $3.6 \times 4.8 = 17.4$, was very modest. The huge advantage of electrons as a microscope radiation source, compared to light, is that they can be given energies that correspond to very short wavelengths. Thus the theoretical resolving power of electron microscopy is very high. Lenses cannot yet be made satisfactorily for other short wavelength radiation sources, such as X-rays.

With some delays because of extraneous events, like World War II, electron microscopes were developed commercially by Siemens and others in the 1940s. Electron microscopy was the first tool capable of giving images of biological materials at a macromolecular level. The problem is that the sample is bombarded by electrons in a vacuum, and the contrast produced by most biological samples is very low. A typical early example was the observation of virus particles by Brenner and Horne (1959). Sample protection and good contrast were achieved by mixing the virus preparations with 1% phosphotungstic acid and spraying onto a supporting grid made from carbon film.

Images improved with better instrumentation over the years, but further important advances were made in the 1960s—averaging and image processing techniques were introduced that began to yield much better images. De Rosier and Klug (1968) were the first to determine a 3D structure from electron images when they applied image reconstruction and computational methods to investigations of the T4 bacteriophage tail (Figure 2).

A major breakthrough came when Henderson and Unwin (1975) determined a low resolution (7 Å) 3D structure of bacteriorhodopsin. They achieved this by a combination of electron diffraction and electron microscopy images of 2D arrays of the protein in the "purple membrane," what is now called *Halobacterium halobium*. Different images were collected, with the specimen tilted at different angles to the electron beam. Good electron diffraction data were collected but, unfortunately, the isomorphous replacement methods with heavy atoms, so successfully used with X-rays, could not be used because of insufficient contrast with electrons. The diffraction data were thus combined with phase information extracted from electron microscopy images to reconstruct the model shown in Figure 3. The protein in the membrane was shown to contain seven α-helices which were roughly perpendicular to the membrane plane. Lipid bilayer regions were found in the spaces between the protein molecules. This was the first information obtained about the structure of an integral membrane protein. This seven-helix protein also turned out to be a member of a very large and important family of transmembrane signaling proteins.

It took Richard Henderson and colleagues another 15 years to significantly improve this model of bacteriorhodopsin using these methods based on electron microscopy (Henderson et al., 1990). A 3.5 Å resolution in the plane of the membrane was achieved, while the resolution out of the plane was not as good because of the problem of sample tilting beyond about ± 60°. Why did this take so long compared to the very rapid advances made by the X-ray crystallographers studying 3D protein crystals? Several problems with this kind of microscopy of 2D crystals had to be overcome. One is the presence of sample imperfections—flat arrays of 2D crystals are very difficult to prepare. There are also several instrumen-

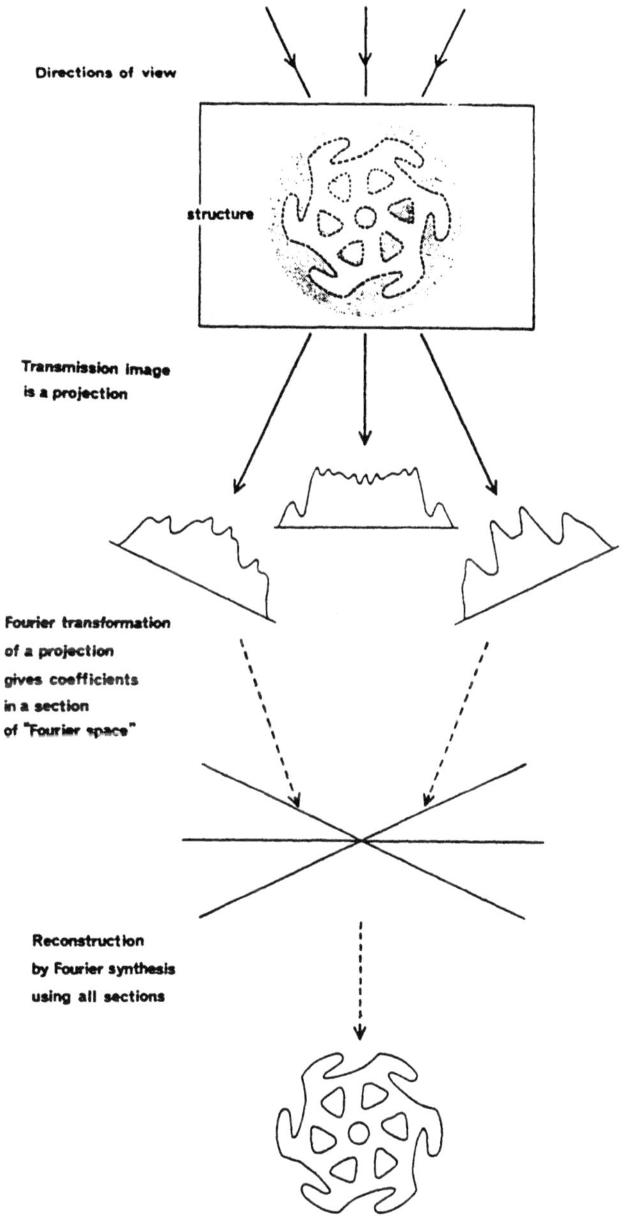

Figure 2. Scheme illustrating how a series of electron micrograph images, taken at different tilt angles, can be reconstructed to yield a 3D structure (de Rosier and Klug, 1968, with permission).

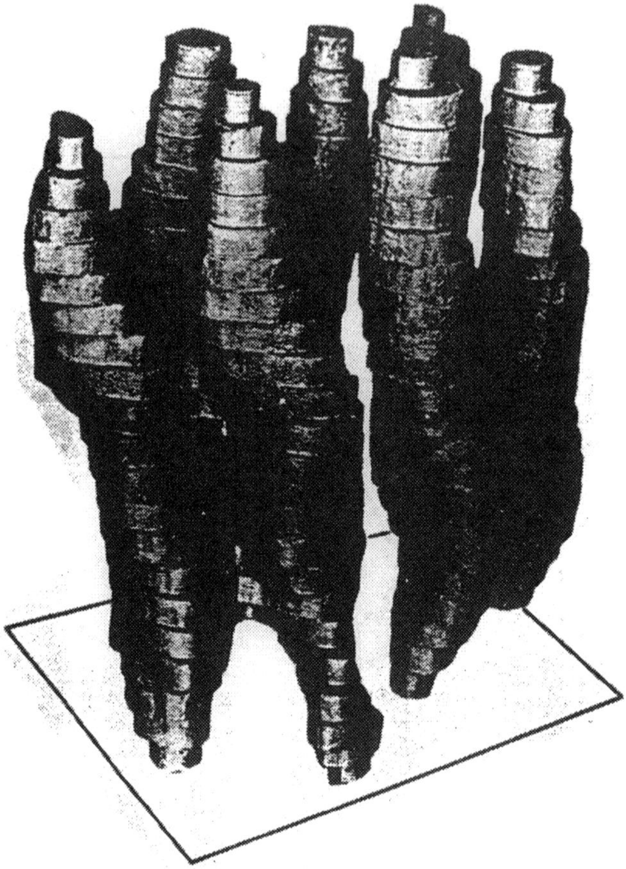

Figure 3. A model of a single molecule of bacteriorhodopsin in the purple membrane, obtained using tilted electron micrographs and electron diffraction. The square represents the plane of the membrane (from Unwin and Henderson, 1975, with permission).

tal difficulties in producing images of high enough quality. This meant that complex algorithms had to be developed to process the images in a way that compensated for these imperfections. The electron beam can also cause radiation damage to the sample and an important instrumental advance was to lower the sample temperature in the electron microscope in order to reduce sample instability.

Nuclear Magnetic Resonance

Two groups independently demonstrated that nuclei, with the property of "spin," could be made to resonate in an applied magnetic field (Bloch et al., 1946; Purcell et al., 1946). This was the birth of a new tool, nuclear magnetic resonance (NMR). It is a very low-energy method, with typical detection wavelengths of around 1 m, yet it is extraordinarily versatile. It can determine the structure of macromolecules in solution (Wüthrich, 1989), measure metabolite concentrations in intact tissue (Radda, 1992), obtain images of biological materials with 10 μm resolution (Bowtell et al., 1992), and, in a relatively new development, make pairwise distance measurements between isoptopically labeled groups in a membrane protein (Smith and Peersen, 1992). Here I will concentrate on macromolecular structure determination in solution.

Initially, the potential of NMR in biology was far from realized. It first came as a surprise that the frequency of the signals, originating from the nucleus, were very sensitive to the nature of the chemical bonding, i.e. the electronic environment of the nuclei (Proctor and Yu, 1950). This property is usually called the "chemical shift." As the homogeneity of available magnets improved, another important phenomenon was observed—the resonances of chemical groups can exhibit fine structure, or spin–spin coupling, that arises from interactions with nuclei sharing bonding electrons (Gutowski et al., 1951). These two phenomena, chemical shift and spin–spin coupling, have the capacity to label a given molecule with a unique and characteristic NMR spectrum. NMR was thus soon recognized as an extremely powerful analytical tool, one which rapidly became established in chemistry.

Within 10 years of the discovery, most of the important features of NMR experiments had been elegantly explained. For example, an early paper (Overhauser, 1953) suggested that the intensity of an NMR signal would be enhanced by irradiation of unpaired electron spins in a sample. This led to a method, involving observation of the "nuclear Overhauser effect" (NOE), where a mutual dipolar interaction between two protons can be detected, provided that the proton dipoles are less than about 5 Å apart. This distance information is the basis for structure determination by NMR. This also explains why a long wavelength technique can give high resolution information; the high resolution arises because signals from individual chemical groups in the macromolecule can be resolved. The main advantage of NMR over other structural methods is that it can be done in an aqueous solution of the macromolecule; the crystals and heavy atom derivatives required by X-ray crystallography are not needed.

The information content in the first published "high resolution" NMR spectrum (see Figure 4) of a protein was very low (Saunders et al., 1957), with no signals from particular groups being resolved. All the tools of structural biology have been improved greatly since their inception, but technical improvements have been particularly striking with NMR. The first published protein spectrum was obtained using a spectrometer operating at 40 MHz [a field strength of 1 Tesla causes protons (^1H) to resonate at around 42.6 MHz]. Spectrometers operating with applied field strengths of over 18 Tesla are now in use all over the world; these new magnets are not only stronger but are much more homogeneous than early magnets (Boyd et al., 1994). Two other important technical advances were the introduction, in the late

1960s, of pulse methods with Fourier transformation (Ernst and Anderson, 1966) and, 10 years later, of two-dimensional techniques (Aue et al., 1976) (an example is shown in Figure 4b). The contribution of Richard Ernst to these new methods earned him a Nobel prize in 1991. These various developments have led to enormous improvements in instrument sensitivity and resolving power since the first experiments 50 years ago (see Figure 4).

As the resolving power of NMR spectrometers improved, it was realized that chemical shift sensitivity was such that resonances from individual amino acids could be resolved in a folded protein. In other words, two amino acids of the same type were resolvable because they were in a unique environment. The problem was: how can resolved resonances be assigned to particular groups? Early attempts

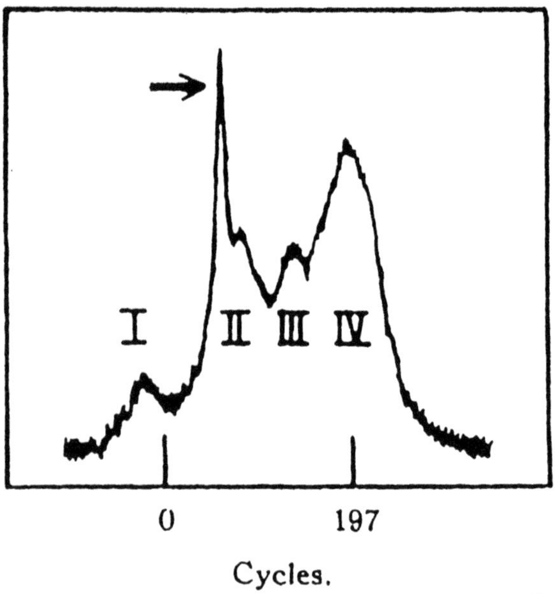

(continued)

Figure 4. (**a**) The first ^{1}H NMR spectrum of a protein, ribonuclease, produced with a spectrometer operating at a frequency of 40 MHz. The arrow identifies the residual solvent water peak. "Because of their location, peaks I and IV can be tentatively assigned to aromatic hydrogens and to hydrogens on aliphatic carbon atoms" (from Saunders et al., 1957, with permission). (**b**) Part of a 2D NMR spectrum of an ^{15}N-labeled sample of a 174 amino acid cytokine, granulocyte colony stimulating factor, taken at 750 MHz. This spectral region shows the NH amide region, with ^{15}N along one axis and ^{1}H along the other. All amino acids, except prolines, have a resonance in this region. The assignment of all the peaks to specific amino acids is shown except for the 33 leucines which are only shown as arrows for space reasons—the leucine assignments are, however, also known (adapted from Werner et al., 1994).

included assignment of side-chain proton resonances of the four histidines in ribonuclease (Meadows et al., 1967) and some unusually shifted methyl resonances in lysozyme (McDonald and Phillips, 1967). As recently as 1970, before very high fields and 2D NMR, it was an impossible task to assign more than a handful of resonances, even for a small 10-kDa protein. As instruments improved, more resonances could be assigned, and there were some attempts to derive systematic

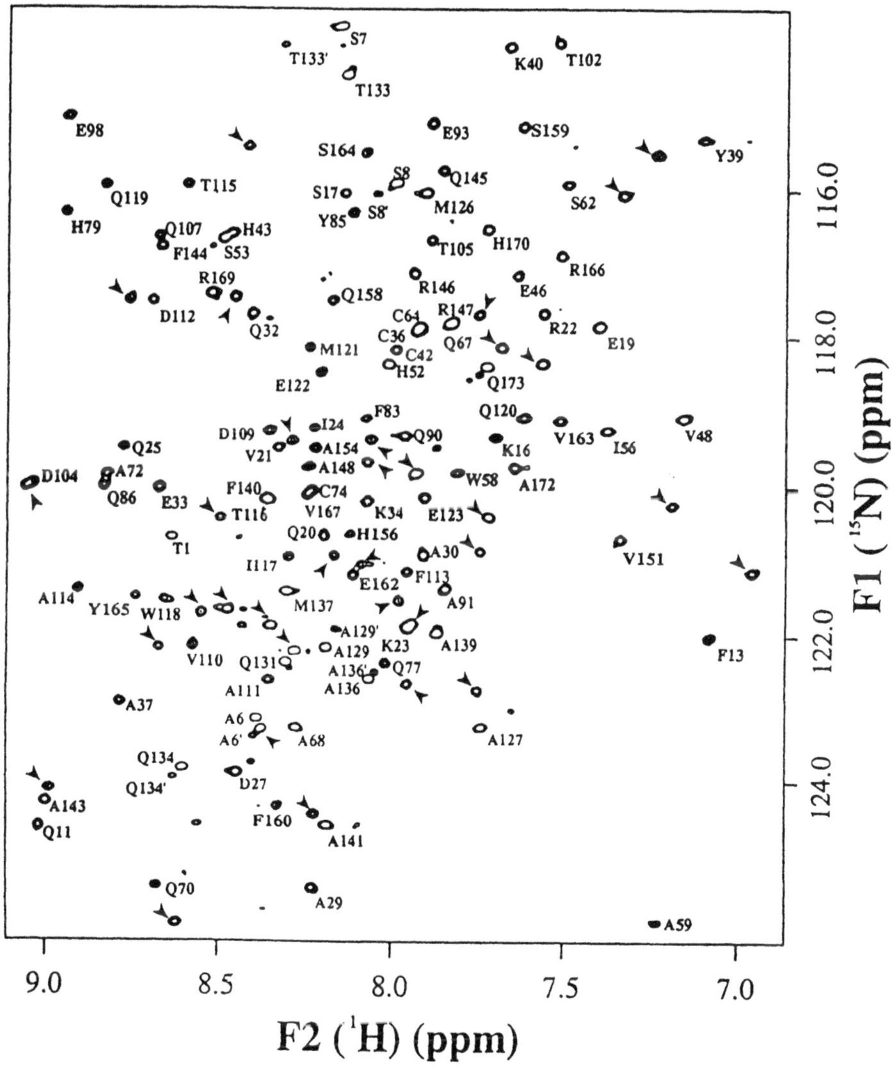

Figure 4. (Continued)

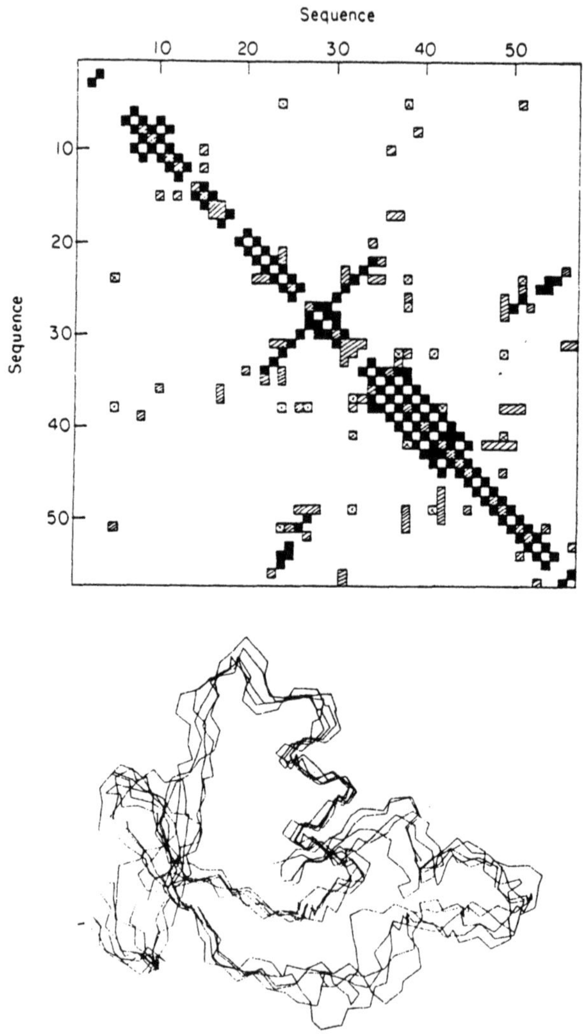

Figure 5. The first calculation of a protein structure from NMR data (Williamson et al., 1985). The protein is a 57 residue proteinase inhibitor (BUSI IIA) dissolved in water. (**a**) Diagonal plot of some of the observed dipole–dipole interactions (NOEs) between pairs of protons. Each square indicates the presence of a through-space connection. The region around residues 34–45, where there are many connections near the diagonal, corresponds to an α-helix; β-sheet regions can be identified by connectivities running perpendicular to the diagonal, e.g. the residues 27–29 form a sheet with residues 47–49. (**b**) If the restraints of the kind illustrated in (**a**) are used in computer simulations, families of structures that are consistent with these restraints can be calculated. This diagram shows the backbone atoms of five such structures overlaid.

structural information from proteins; an example is mapping the distances of certain amino acid groups from a paramagnetic metal ion in lysozyme (Campbell et al., 1975). It is important to realize, however, that in these early examples many of the resonance assignments depended on prior structural knowledge.

The first systematic NMR investigations of protein structure came from Kurt Wüthrich and colleagues. They demonstrated how spectral assignment could be done by using the pattern of NOE effects observed between amino acid neighbors (Wagner and Wüthrich, 1982). This information could be coupled with knowledge of the primary sequence to produce sequence specific assignments independent of 3D structure information. Procedures for calculating structures from the many pairwise distances that could be deduced from the NMR data were also laid down at this time (Havel and Wüthrich, 1985). This led to the first complete NMR structure determination of a protein in aqueous solution, as illustrated in Figure 5 (Williamson et al., 1985). [Similar calculations were also carried out earlier by the same group on glucagon bound to a micelle system (Braun et al., 1983).]

These papers, published only 10 years ago, have led to an explosion of structure determinations using NMR techniques. This has been very valuable especially for aqueous solutions of proteins with mobile regions that do not crystallize readily. It is, however, important to realize that the information content available from NMR studies of proteins in solution depends on the molecular weight of the protein. With increasing mass, the spectral lines become broader and the overlap of resonances becomes more severe. This means that resolution and assignment of individual resonances become very difficult with proteins larger than about 25 kDa.

THE CURRENT STATUS OF STRUCTURAL BIOLOGY (TODAY)

Biochemistry in the late twentieth century is characterized by many excellent text books, beautifully illustrated with numerous detailed color graphics pictures of macromolecular structures. These pictures are used for several reasons; one, no doubt, is to promote sales, but they also have great artistic appeal and give considerable insight into the workings and beauty of nature. I think few, layman or biochemist, could fail to be impressed by the spectacular new structures produced within the last few years. Most of these have been produced by X-ray diffraction methods, but low temperature electron microscopy (cryo-EM) and NMR are making an increasingly important impact. New structures are currently produced at a rate of about one per day and are deposited in the rapidly growing Brookhaven Protein Database (PDB). In addition there are numerous new scientific journals completely devoted to structural biology, e.g. *Nature, Structural Biology,* and *Structure*. All this is a long way from the painstaking production of the "visceral object" view of myoglobin in 1958 (Figure 1).

Improvements in the technology of the three main methods of structural biology—X-ray diffraction, electron microscopy, and NMR—have been remarkable over the last decade. Astonishing advances in computational power have been a major factor for all the methods. A very important additional component of modern advances in structural biology is the ability to express any specified region of a protein by recombinant methods and to manipulate the composition of this protein. Examples are the incorporation of seleno-methionine to help solve the phase problem in diffraction studies and isotopic labeling for NMR.

New radiation sources are having a major impact on X-ray diffraction methods. New synchrotrons provide irradiation that is several orders of magnitude brighter than just a few years ago (Brändén, 1994). New detection devices, such as image plates, collect the diffraction data much more readily and efficiently than photographic film. In addition, powerful computational methods have greatly facilitated the critical steps of phase determination and structure refinement. Refinement is where a model structure, built on the basis of an electron density map, is improved by a process of cycling between diffraction pattern and the model structure. This is an important step since initial models may be far from correct because of uncertainties in the electron density map (see e.g. Brändén and Jones, 1990).

High resolution studies with electron diffraction/microscopy now use very low temperatures (cryo-EM) routinely. Microscopes continue to improve, and exploitation of lower sample temperatures and scanning methods are giving better resolution. The possibility of having low temperature samples has been used recently to trap membrane proteins in different states. Subramanian et al. (1993) observed structural changes at 3.5 Å resolution in bacteriorhodopsin during the photocycle. Nigel Unwin (1995), in a spectacular and now unusual single-author work, was able to observe the acetylcholine receptor in the open state at 9.5 Å using similar methods. He was thus able to complement his earlier studies of acetylcholine receptor in the closed state. Numerous other exciting structures are emerging using this electron/2D crystal methodology—for example, the light-harvesting chlorophyll *a*/*b* complex (Kühlbrandt et al., 1994). These studies are shedding light on an area that has been particularly difficult for structural biology—that of membrane proteins. Until recently there was concern that the growing database of solved structures was seriously skewed towards soluble water soluble proteins.

One of the more difficult areas of structural biology is to solve the structure of macromolecular complexes. One reason for this is that it is extremely difficult to obtain crystals of such systems; another is that the complexity of the problem becomes unmanageable. Success has been obtained with large virus structures partly because they are highly symmetric. Other systems, like the ribosome, have been studied for a long time and progress is being made, but rather slowly (Yonath and Franceschi, 1993). One way of tackling the "large system" problem is to "divide and conquer." Like many of the ideas of structural biology, this was first advanced in the Laboratory of Molecular Biology (LMB) in Cambridge (Klug, 1978). Structures of component parts of a large macromolecule, or a complex, can be solved to high resolution, and these components can then be fitted to lower resolution pictures of the intact systems. A recent and beautiful example of this approach is the combination of high resolution X-ray diffraction studies of actin monomers and the myosin head with cryo-EM views of decorated actin to produce a high resolution model of a major component of muscle (Rayment et al., 1993).

NMR gives structural information without a requirement for crystals or heavy atoms, but the method is restricted to small proteins. Some improvement comes from the continuing implementation of better spectrometers, but other ways of alleviating the overlap problem, such as using isotopic labels and multidimensional pulse techniques, have also been devised. Protein produced by recombinant methods is also relatively easy to label with ^{13}C and ^{15}N. These tricks have made it possible to assign completely the spectra of proteins up to about 30 kda (see e.g. Fogh et al., 1995); it seems unlikely, however, that this limit will be greatly extended in the future. Because of the size problem, the "divide and conquer" or "dissection" approach mentioned above for complexes is particularly important when applying NMR to structural problems. NMR can, however, also yield a wide range of information about proteins that complements straightforward structural data, including data on protein dynamics, the nature of protein folding intermediates, enzyme kinetics, and the ionization states of individual side chains.

Recent examples of the application of NMR to protein structure are the demonstration a major rearrangement of the two calcium binding regions in calmodulin when it binds to a peptide (Ikura et al., 1992), and evidence for a calcium-induced conformational change (see e.g. Finn and Forsen, 1995). Another powerful use of NMR is the application of the dissection approach to multidomain proteins. Analysis of protein sequences has shown that many proteins, especially from multicellular organisms, appear to be made up from a limited number of autonomously folding protein domains or modules that are usually less than 200 amino acids in length (Baron et al., 1991). These modules appear numerous times in different proteins—for example, the epidermal growth factor like module has been found about 1000 times in the sequence database. Such modules are, fortunately, within the size range of the NMR method. Numerous single- and double-module structures, dissected from modular proteins associated with cell adhesion and signaling events, have now been solved (see e.g. Campbell and Downing, 1994).

The light microscope and its wavelength limitations have been mentioned above. Other microscopes are being developed that do not depend on wavelength—for example, "atomic force" microscopes that essentially "feel" the shape of a sample, mechanically. A recent interesting example of an application of atomic force microscopy is a study of lysozyme in the presence and absence of substrate (Radmacher at al., 1994). While that study indicates some of the potential uses of these new methods, the complete determination of a 3D protein structure by this means still seems a long way off. Some information about a structure can, of course, also be extracted from various spectroscopic methods other than NMR (see recent reminiscences on developments of spectroscopy methods by Beinert, 1994 and Udenfriend, 1995), but while these methods can give information about local geometry and overall shape, they do not give information about 3D folds.

Further information about the current status of the tools of structural biology can be found in recent reviews. Electron microscopy is covered by Chiu (1993) and de Rosier (1993); the methodology of NMR structure determination has been reviewed extensively (Wüthrich, 1989; Clore & Gronenborn, 1992; Evans, 1995), and a good recent book on X-ray diffraction methods has been written by Drenth (1994).

FUTURE PROSPECTS FOR STRUCTURAL BIOLOGY (TOMORROW)

What about risking predictions for the future for structural biology? There are about 100,000 genes in the human genome whose sequence will soon be known. There are, almost certainly, not 100,000 completely different protein structures corresponding to these genes; they will exist as a limited subset of protein structural types. Cyrus Chothia (1992) predicted that there would only be about 1000 protein families and that outline structures would be available for all of them in time for the completion of the Human Genome Project by about 2015. If the human genome is sequenced and all the basic structural motifs are known, what will be left for structural biologists to do? Fortunately (for those of us in the trade!), there will still be much to be done. Many of the large intricate macromolecular complexes in cells are still beyond the scope of even the most powerful modern structural tools. The "divide and conquer" strategy will thus continue to be heavily used. Even with knowledge of structural components, there will still be a need to gather information about how the various structural motifs move and fit together in a functioning cell.

What are the prospects for new tools for structural biology being developed? No doubt all the current methods will continue to improve with higher sensitivity, more stability, and higher resolution. Computational power will continue to increase, thus allowing unprecedented modeling and simulation calculations. Structure prediction tools will be greatly enhanced not only because of increased understanding of the way proteins fold but also from comparisons within a large database of structural knowledge. It also seems clear that various forms of scanning microscopy, including light-based methods (Paddock, 1994), will contribute increasingly to our overall picture of functioning molecules in living cells.

CONCLUSIONS

I have tried to indicate the scope of the astonishing developments in structural biology over the last 50 years, and their origins. The current exciting state of affairs has partly come about by remarkable improvements in the sophisticated and expensive tools of the trade. A satisfying feature of the three main tools currently available is that the information they yield about molecules is complementary. Diffraction methods, applied to 3D crystals, are the most powerful structural tool but they do require crystals. NMR has a serious size limitation, but it does work in solution. Cryo-EM methods require ordered arrays of molecules in relatively harsh conditions, but they are often the only way to look at membrane proteins. As well as powerful hardware, other factors have played an important role in the achievements of structural biology; not least are advances in chemical knowledge about biological macromolecules and the ability to manipulate and produce them. This powerful combination of different tools means that we are now poised to produce more information about the structure and function of biological systems than ever before. Perhaps the "breakthrough" stage is over; we are now well into "consolidation" where the accumulating data will help us to have a better understanding of the workings of the natural world. We are also beginning, under increasing Government and financial pressure, to accelerate the "exploitation" phase of the campaign where structural knowledge will be used to contribute to our health and wealth.

ACKNOWLEDGMENTS

This is a contribution from the Oxford Centre for Molecular Science which is supported by BBSRC and MRC. I also gratefully acknowledge support from the Wellcome Trust.

REFERENCES

Aue, W.P., Bartholdi, E., & Ernst, R.R. (1976). Two-dimensional spectroscopy. Application to nuclear magnetic resonance. J. Chem. Phys. 64, 2229–2246.

Baron, M., Norman, D., & Campbell, I.D. (1991). Protein modules. TIBS 16, 13–17.

Beinert, H. (1994). Looking at enzymes in action in the 1950s. Protein Science 3, 1605–1612.

Bloch, F., Hansen, W.W., & Packard, M. (1946). Nuclear induction. Phys. Rev. 69, 127.

Bowtell, R.W., Brown, P.M., Glover, P.M., McJury, M., & Mansfield, P. (1990). Resolution of cellular structures by NMR microscopy at 11.7 T. Phil. Trans. R. Soc. Lond. A 333, 457–467.

Boyd, J., Soffe, N., & Campbell, I.D. (1994). NMR at very high fields. Structure 2, 253–255.

Bracegirdle, B. (1989). Microscopy and comprehension: the development of understanding of the nature of the cell. TIBS 14, 464–468.

Bragg, L. (1967). Introduction ("A discussion on the structure and function of lysozyme" organised by Perutz). Proc. Roy. Soc. Lond. Ser. B 167, 349.

Brändén, C.-I. (1994). The new generation of synchrotron machines. Structure 2, 5–6.

Brändén, C.-I. & Jones, T.A. (1990). Between objectivity and subjectivity. Nature 343, 687–689.

Braun, W., Wider, G., Lee, K.H., & Wüthrich, K. (1983). Conformation of glucagon in a lipid–water interphase by ^{1}H nuclear magnetic resonance. J. Mol. Biol. 169, 921–948.

Brenner, S. & Horne, R.W. (1959). A negative staining method for high resolution electron microscopy of viruses. Biochim. Biophys. Acta 34, 103–110.

Busch, H. (1927). Über die Wirkungsweise der Konzentrierungsspule bei der Braunschen Röhre. (On the mode of action of the concentrating coil in the Braun tube.). Arch. Elektrotechnil 18, 583–594.

Campbell, I.D., Dobson, C.M., & Williams, R.J.P. (1975). Nuclear magnetic resonance studies of the structure of lysozyme in solution. Proc. Roy. Soc. Lond. A. 345, 41–59.

Campbell, I.D. & Downing, A.K. (1994). Building protein structure and function from modular units. TIBTECH 12, 168–172.

Chiu, W. (1993). What does electron cryomicroscopy provide that X-ray crystallography and NMR spectroscopy cannot? Annu. Rev. Biophys. Biomol. Struct. 22, 233–255.

Chothia, C. (1992). One thousand families for the molecular biologist. Nature 357, 543–544.

de Rosier, D.J. (1993). Turn-of-the-century electron microscopy. Current Biology 3, 690–692.
de Rosier, D.J. & Klug, A. (1968). Reconstruction of three dimensional structures from electron micrographs. Nature 217, 130–134.
Drenth, J. (1994). Principles of Protein X-Ray Crystallography. Springer-Verlag, New York.
Eisenberg, D. (1994). Max Perutz's achievements: How did he do it? Protein Science 3, 1625–1628.
Ernst, R.R. & Anderson, W.A. (1966). Application of Fourier transform to magnetic resonance. Rev. Sci. Inst. 37, 93–102.
Evans, J.N.S. (1995). Biomolecular NMR Spectroscopy. Oxford University Press, New York.
Finn, B.E. & Forsén, S. (1995). The evolving model of calmodulin structure, function and activation. Structure 3(1), 7–11.
Fogh, R.H., Schipper, D., Boelens, R., & Kaptein, R. (1995). Complete ^{1}H, ^{13}C and ^{15}N NMR assignments and secondary structure of the 269-residue serine protease PB92 from *Bacillus alcalophilus*. J. Biomolecular NMR 5, 259–270.
Gutowski, H.S., McCall, D.W., & Schlichter, C.P. (1951). Coupling among nuclear magnetic dipoles in molecules. Phys. Rev. 84, 589–590.
Havel, T.F. & Wüthrich, K. (1985). An evaluation of the combined use of nuclear magnetic resonance and distance geometry for the determination of protein conformations in solution. J. Mol. Biol. 185, 281–294.
Henderson, R., Baldwin, J.M., Ceska, T.A., Zemlin, F., Beckmann, E., & Downing, K.H. (1990). Model for the structure of bacteriorhodopsin based on high-resolution electron cryomicroscopy. J. Mol. Biol. 213, 899–929.
Henderson, R. & Unwin, P.N.T. (1975). Three-dimensional model of purple membrane obtained by electron microscopy. Nature 257, 28–32.
Ikura, M., Clore, G.M., Gronenborn, A.M., Zhu, G., Klee, C.B., & Bax, A. (1992). Solution structure of a calmodulin-target peptide complex by multidimensional NMR. Science 256, 632–638.
Clore, G.M. & Gronenborn, A.M. (1992). Structures of larger proteins in solution: three and four dimensional heteronuclear NMR spectroscopy. Science 252, 1390–1399.
Kendrew, J.C., Bodo, G., Dintzis, H.M., Parrish, R.G., Wyckoff, H., & Phillips, D.C. (1958). A three-dimensional model of the myoglobin molecule obtained by X-ray analysis. Nature 181, 662–666.
Klug, A. (1978–79). Image analysis and reconstruction in the electron microscopy of biological macromolecules. Chemica Scripta 14, 245–256.
Kossiakoff, A.A. (1985). The application of neutron crystallography to the study of dynamic and hydration properties of proteins. Ann. Rev. Biochem. 54, 1195–1227.
Kühlbrandt, W., Wang, D.N., & Fuyiyoshi, Y. (1994). Atomic model of plant light-harvesting complex by electron crystallography. Nature 367, 614–621.
McDonald, C.C. & Phillips, W.D. (1967). Manifestations of the tertiary structures of proteins in high-frequency nuclear magnetic resonance. J. Am. Chem. Soc. 89, 6332–6341.
Meadows, D.H., Markley, J.L., Cohen, J.S., & Jardetzky, O. (1967). Nuclear magnetic resonance studies of the structure and binding sites of enzymes. I. Histidine residues. Proc. Natl. Acad. Sci. USA 58, 1307–1313.
Overhauser, A. (1953). Polarization of nuclei in metals. Phys. Rev. 92, 411–415.
Paddock, S.W. (1994). To boldly glow . . . . . Applications of laser scanning confocal microscopy in developmental biology. BioEssays 16, 357–365.
Perutz, M. (1985). Early days of protein crystallography. Methods in Enzymology 114, 3–19.
Perutz, M. (1992). Protein structure: new approaches to disease and therapy. W.H. Freeman, New York.
Perutz, M.F. (1964). The Hemoglobin Molecule. Readings from *Scientific American*. The Molecular Basis of Life. W.H. Freeman, New York.
Phillips, D.C. (1971). Protein crystallography 1971: coming of age. Cold Spring Harbor Symp. Quant. Biol. 36, 589–592.

Proctor, W.G. & Yu, F.C. (1950). The dependence of a nuclear magnetic resonance frequency upon chemical compound. Phys. Rev. 77, 717.

Purcell, E.M., Torrey, H.C., & Pound, R.V. (1946). Resonance absorption by nuclear magnetic moments in a solid. Phys. Rev. 69, 37.

Radda, G.K. (1992). Control, bioenergetics and adaption in health and disease: noninvasive biochemistry from nuclear magnetic resonance. FASEB J. 6, 3033–3038.

Radmacher, M., Fritz, M., Hansma, H.G., & Hansma, P.K. (1994). Direct observation of enzyme activity with the atomic force microscope. Science 265, 1577–1579.

Rayment, I., Holden, H.M., Whittaker, M., Yohn, C.B., Lorenz, M., Holmes, K.C., & Milligan, R.A. (1993). Structure of the actin–myosin complex and its implications for muscle contraction. Science 261, 58–65.

Rossmann, M.G. (1994). The beginnings of structural biology. Protein Science 3, 1731–1733.

Ruska, E. (1987). The development of the electron microscope and of electron microscopy. Bioscience Reports 7(8), 607–629.

Saunders, M., Wishnia, A., & Kirkwood, J.G. (1957). The nuclear magnetic resonance spectrum of ribonuclease. J. Am. Chem. Soc. 79, 3289–3290.

Schoenborn, B.P. (1969). Neutron diffraction analysis of myoglobin. Nature 224, 143–146.

Smith, S.O. & Peersen, O.B. (1992). Solid-state NMR approaches for studying membrane protein structure. Ann. Rev. Biophys. Biomol. Struct. 21, 25–47.

Stryer, L. (1968). Implications of X-ray crystallographic studies of protein structure. Ann. Rev. Bioch. 37, 25–50.

Subramaniam, S., Gerstein, G., Oesterhelt, D., & Henderson, R. (1993). Electron diffraction analysis of structural changes in the photocycle of bacteriorhodopsin. EMBO 12(1), 1–8.

Udenfriend, S. (1995). Development of the spectrophotofluorometer and its commercialization. Protein Science 4, 542–551.

Unwin, N. (1995). Acetycholine receptor channel imaged in the open state. Nature 373, 37–43.

Wagner, G. & Wüthrich, K. (1982). Sequential resonance assignments in protein ^{1}H NMR spectra: basic pancreatic trypsin inhibitor. J. Mol. Biol. 155, 347.

Werner, J., Breeze, A., Kara, B., Rosenbrock, G., Boyd, J., Soffe, N., & Campbell, I.D. (1994). Secondary structure and backbone dynamics of human granulocyte colony stimulating factor in solution. Biochemistry 33, 7184–7192.

Williamson, M.P., Havel, T.F., & Wüthrich, K. (1985). Solution conformation of proteinase inhibitor IIA from bull seminal plasma by ^{1}H nuclear magnetic resonance and distance geometry. J. Mol. Biol. 182, 295–315.

Wüthrich, K. (1989). Protein structure determination in solution by nuclear magnetic resonance spectroscopy. Science 243, 45–50.

Yonath, A. & Franceschi, F. (1993). Structural aspects of ribosomes. Current Opinion in Structural Biology 3, 45–49.

Chapter 7

GLYCOBIOLOGY:
A QUANTUM LEAP IN CARBOHYDRATE CHEMISTRY

R.A. Dwek

INTRODUCTION AND BACKGROUND

The chemistry of simple sugars was worked out in the late nineteenth century by Emil Fischer, and the ring structures determined in the inter-war years by Haworth and colleagues. Simple polysaccharides such as starch, glycogen, and cellulose, as well as more complex molecules such as chitin and hyaluronic acid had also received attention and their component sugars identified by classical means. By the 1960s, especially through work on blood-group determinants by Morgan, Watkins, and their associates at the Lister Institute, it had become clear that besides simple mono- and polysaccharides, naturally occurring carbohydrates were commonly conjugated to proteins and lipids (as glycoproteins and glycolipids). Mucopolysaccharides and proteoglycans were also described, distinguishable by their relative proportions of carbohydrate to protein, but little progress could be made to determine the structure or function of these complex molecules until sensitive and sophisticated techniques became available to analyze the component sugars and the order and structural details of their attachment to protein. Now automatic techniques are available for analysis of glycoproteins.

The three major classes of macromolecules in biology are DNA, proteins, and carbohydrates. Proteins and nucleic acids are almost exclusively linear and they have only a single type of linkage between amide bonds for proteins and 3′-5′-phosphodiester bonds for nucleic acids. Carbohydrates differ from the other two classes of biological polymers in two important characteristics: they can be highly branched molecules, and their monomeric units may be connected to one another by many different linkage types. This complexity allows carbohydrates to provide almost unlimited variations in their structures; a linear sequence of three different monosaccharide units can be chemically linked together in more than 35,000 ways. Although carbohydrates can be present without being attached to other molecules, the majority of carbohydrates present in cells are attached to proteins or lipids and the terminology, glycoprotein and glycolipids, is used to reflect this. Further, the attached carbohydrate is often referred to as an oligosaccharide (*Greek Oligo meaning a few*). Glycobiology is the term coined (Dwek, 1988) to describe the expanding roles of glycoconjugates in the function of biological systems.

Glycoproteins are now known to be fundamental to many important biological processes including fertilization, immune defense, viral replication, parasitic infection, cell growth, cell–cell adhesion, degradation of blood clots, and inflammation. They are major components of the outer surface of mammalian cells. Over half the biologically important proteins are glycosylated. Oligosaccharide structures change dramatically during development and it has been shown that specific sets of oligosaccharides are expressed at distinct stages of differentiation. Further, alterations in cell surface oligosaccharides are associated with various pathological conditions including malignant transformation.

ANALYTICAL PROCEDURES

In the early 1980s the determination of carbohydrate sequences was very difficult and was carried out by very few laboratories, mainly in Japan by Akira Kobata and colleagues. As in all newly developing fields, technology played a crucial role. In 1983, the Monsanto Company entered into a close partnership with Oxford University to develop jointly the technology necessary to release, isolate, and sequence oligosaccharides from glycoconjugates in nanomole amounts. Figure 1 shows a schematic of a complete process for characterizing glycoprotein glycans.

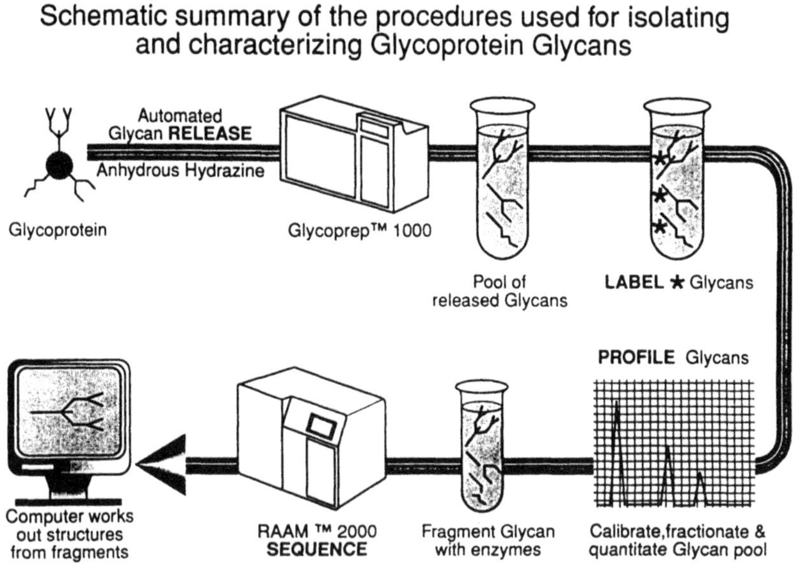

Figure 1. An overall process for release, labeling, and sequencing of nanomolar amounts of a glycan from a glycoprotein.

Release of Glycans from Glycoprotein

In order to release glycans, a *general* method is required that is independent of the conjugate to which the glycan is attached. For this reason a chemical method is used, namely the use of hydrazine to cleave both N- and O-glycosidic bonds. Hydrazinolysis releases glycans which are intact and have a free reducing terminus, is nonselective (with respect to the glycan), and the process is amenable to automation. Released glycans are separated from peptide fragments, leaving unreduced, intact glycans ready for labeling and analysis. Enzymatic methods are also extensively used to release glycans from peptides and denatured glycoproteins. However, care must be taken to ensure that the release is nonselective.

Labeling of Released Glycans to Enable their Detection in Subsequent Procedures

This involves a reaction of the reducing terminus of individual glycans using a method which must be independent of the glycan sequence. Two methods which are commonly used are reductive amination with a fluorescent compound, such as 2-aminobenzamide, and reduction with alkaline sodium borotritide, NaB^3H_4, to give the radiolabeled derivative.

Profiling the Pool Glycans to Determine the Types of Glycans Present and their Relative Molar Properties

Three types of glycan profiling are commonly used:

Mass Profile. The molecular weight of each glycan present can be quickly determined by mass spectrometry. Methods which give only the parent ion for each glycan include matrix assisted laser desorption ionization (MALDI) (Figure 2a), and produce spectra on neutral glycans which are fairly easy to interpret, although laser energy induced desialylation can often be observed. Detection limits are in the region of 1 pmol for a single oligosaccharide.

Size Profile. Gel permeation chromatography (GPC) is frequently used to determine the size profile (in glucose units G.U.) of a deacidified glycan pool by coinjection of dextran hydrolysate standard "ladder" with the sample.

Charge Profile. Anion exchange chromatography (AEC) is used to determine the charge profile of a glycan library.

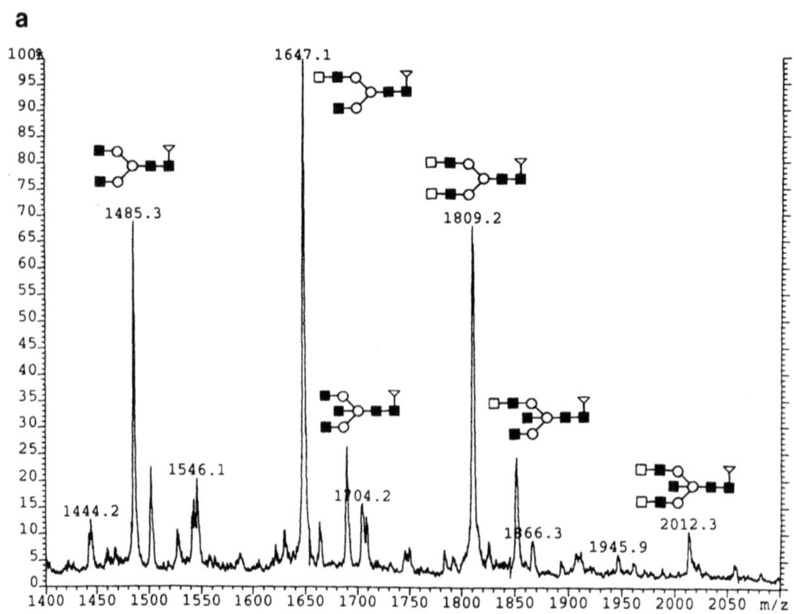

Figure 2. (**a**) Matrix assisted mass desorption spectrum of IgG glycans (*continued*).

Structural Analysis

There is no single technique that is able routinely to provide all the information required for structural analysis. This usually involves the combined use of several physical, chemical, and biochemical techniques including NMR and mass spectrometric and enzymatic analysis. These techniques have been critically reviewed elsewhere (Dwek et al., 1993b). The enzymatic is clearly the method of choice in biological systems for which often only very small amounts of material (picomoles or less) are available.

The basis of enzymatic sequencing for the elucidation of the structure of N- and O-linked glycans is to evaluate the susceptibility of the glycan to a series of sequence-grade exoglycosidases of defined specificity. A recent process for the primary sequence analysis of N-linked oligosaccharides uses exoglycosidases in multiple-defined mixtures, with analysis performed in a single chromatographic step. This process, the reagent array analysis method (RAAM), summarized in Figures 2b and 2c, involves dividing a purified, labeled N-glycan sample into 9 equal aliquots. Each aliquot is incubated with a precisely defined mixture of exoglycosidases called a reagent array. The products of each incubation are combined and a single analysis is performed on the pool of products.

In essence, a mixture of exoglycosidases is used to digest the sample glycan until a linkage is reached which is resistant to all the exoglycosidases present in that mix. By omitting one or more different exoglycosidase(s) from each mixture, different "stop point" fragments of the oligosaccharide are generated. By labeling the

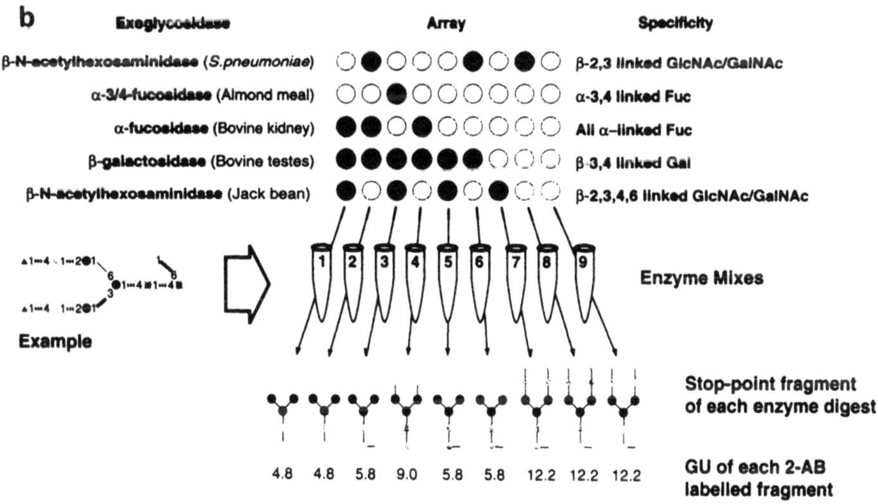

Figure 2. (**b**) The RAAM Enzyme Array consisting of eight different enzyme mixtures and an enzyme bank. (**c**) Summary of oligosaccharide sequencing by RAAM. (*continued*)

original oligosaccharide at the single reducing terminus, fragments retaining the original reducing terminus are readily distinguished from released monosaccharides.

Chromatographic separation of the combined stop point fragments generates a pattern that is, in effect a "signature" of that oligosaccharide treated by the enzyme array used. This signature is characterized by the size (GU) and relative signal intensity of each fragment. A RAAM computer program constructs the carbohydrate structure from the observed signature. The whole process can be accomplished in one or two days, compared with up to a year before the instrumentation was available.

By these means oligosaccharide sequencing is now available routinely.

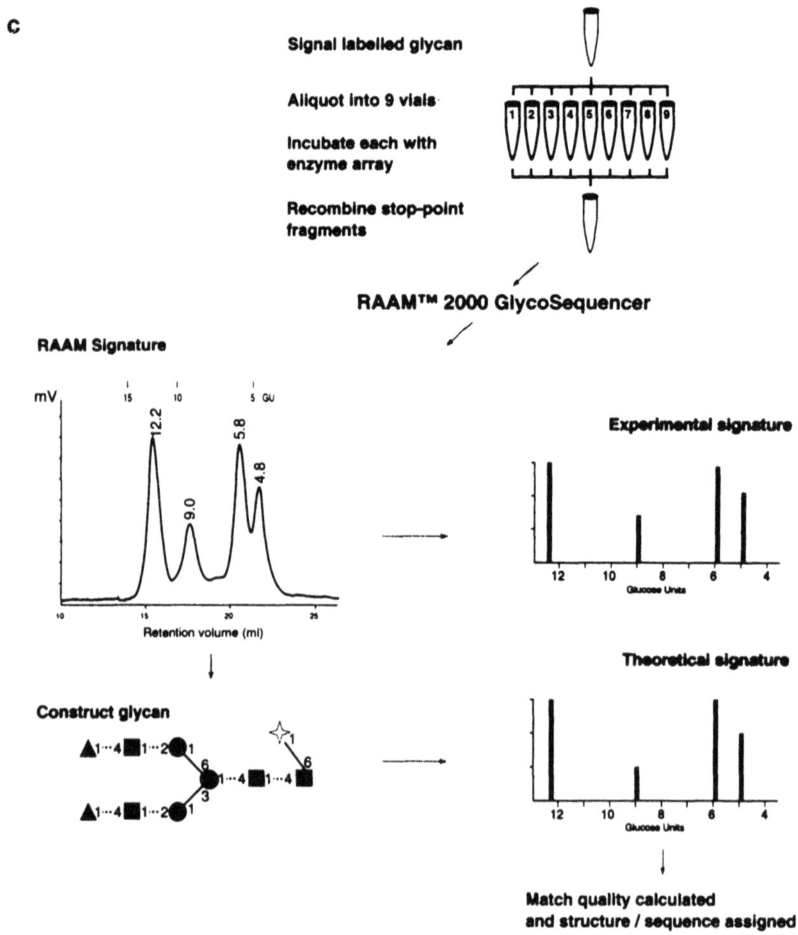

Figure 2. (*continued*)

WHAT DOES A TYPICAL GLYCOPROTEIN LOOK LIKE?

Implication of the Conformations and Dynamics of Protein Surface Oligosaccharides in Protein Function

The majority of cell-surface and secreted proteins are glycosylated, with carbohydrates covalently attached through either a nitrogen atom (supplied by the amino acid asparagine) or an oxygen atom (supplied by serine or threonine). The carbohydrate moiety of a glycoprotein may participate directly in recognition events and may alter the biological function of the protein (Rademacher et al., 1988a; Varki, 1993). A significant factor in modifying the properties of the proteins to which they are attached is postulated to be the large size of the carbohydrates (Parekh et al., 1985, 1989). When the dynamic motions of the carbohydrate are taken into account it becomes apparent that large areas of the protein surface may be shielded by a relatively small oligosaccharide. Further, because of the rigidity of the carbohydrate core, comparatively small motions of the protein–carbohydrate linkage will amplify the motion of the terminal arms of the oligosaccharide. This enables the carbohydrate to span an even larger area of the protein and may have a dramatic effect on the accessibility of the protein in intermolecular interactions.

Accurate quantitation of these properties necessitates a knowledge of both the three-dimensional structures of the carbohydrate and protein and, more importantly, their dynamic behavior. For many proteins conformational information may be obtained from crystallographic methods. Usually oligosaccharides present on glycoproteins appear much less amenable to these techniques, but there are a few examples in which the core residues are clearly defined. These are those in the lectin *Erythrina corallodendron* (Shaanan et al., 1991), the serine protease human leukocyte elastase (HLE) (Bode et al., 1989), the Fc domain of human IgG_1 (Marquart, 1980), and a variant surface glycoprotein from *Trypanosoma brucei* (Freymann et al., 1990). Although the oligosaccharides present in the crystal structures varied in sequence and by the presence or absence of a fucose residue attached to the first GlcNAc residue, the core conformations were remarkably similar to each other. These similarities lead to the conclusion that the conformation of the di-*N*-acetylchitobiose core in N-linked glycoproteins is independent of the protein and would be that present in a solution of the free sugar.

Usually the carbohydrate exhibits greater dynamic fluctuations than the protein. NMR spectroscopy offers insight into these dynamics but NMR data alone are frequently insufficient to determine uniquely the conformations of the oligosaccharide. As an illustration we shall consider the enzyme bovine pancreatic ribonuclease (RNase) for which both experimental data and molecular and dynamic simulations have been carried out.

RNase

RNase is an example of a protein which exists *in vivo* in both nonglycosylated and glycosylated forms, A and B, respectively. RNase B has a single N-linked glycosylation site at Asn-34 and is one of the simplest glycoproteins. Considerable interest has been focused on the differences in the biological functions and properties between RNase A and B (Rudd et al., 1994c). In bovine pancreatic RNase B, biosynthetic processing of the sugar (see below) is halted at the oligomannose stage giving rise to $Man_{5-9}GlcNAc_2$ glycoforms.

Despite the existence of a well-resolved X-ray crystal structure of RNase B, the poor definition of the electron density associated with the oligosaccharide has prohibited any determination of the sugar conformation (Williams et al., 1987). While NMR spectroscopy has been widely applied in the conformational analysis of proteins, including RNase A (Rico et al., 1989, 1991; Robertson et al., 1989; Santoro et al., 1993), and RNase B (Joao et al., 1992), unambiguous conformational determinations of oligosaccharides are less common because of a characteristic paucity of nuclear Overhauser effects (NOEs) between sugar residues. A computer simulation of the dynamic properties of the oligosaccharide offers an alternative approach to the conformational analysis.

Molecular Dynamics (MD) Simulation of $Man_9GlcNAc_2OH$

The application of MD techniques to proteins is typically part of the refinement protocol in X-ray crystallography (Rao and Teeter, 1993). In contrast, MD simulations of oligosaccharides are often applied in conjunction with NMR refinement. This difference leads to unique requirements for the simulations of oligosaccharides. In order to compare MD-generated data with NMR-derived data, the duration of the simulation should be sufficient to sample adequately the conformational space of the macromolecule. A MD trajectory from an unrestrained simulation that is in agreement with the NMR-derived data provides strong support for both the structure and the computational method. During the simulation of Man-9 the core residues were found to maintain a relatively constant conformation, suggesting that in N-linked glycoproteins the conformation of the oligosaccharide is independent of the protein (Figure 3a).

Outer-Arm Conformation of the Oligosaccharide

The conformations of the remaining glycosidic linkages from the MD simulations are in good agreement with values of the glycosidic angles derived from previous NMR studies (Brisson and Carver, 1983a,b; Homans et al., 1986, 1987; Wooten et al., 1990a,b; Woods et al., 1994b) and from MD simulations of related mannobiosides (Woods et al., 1993, 1994b).

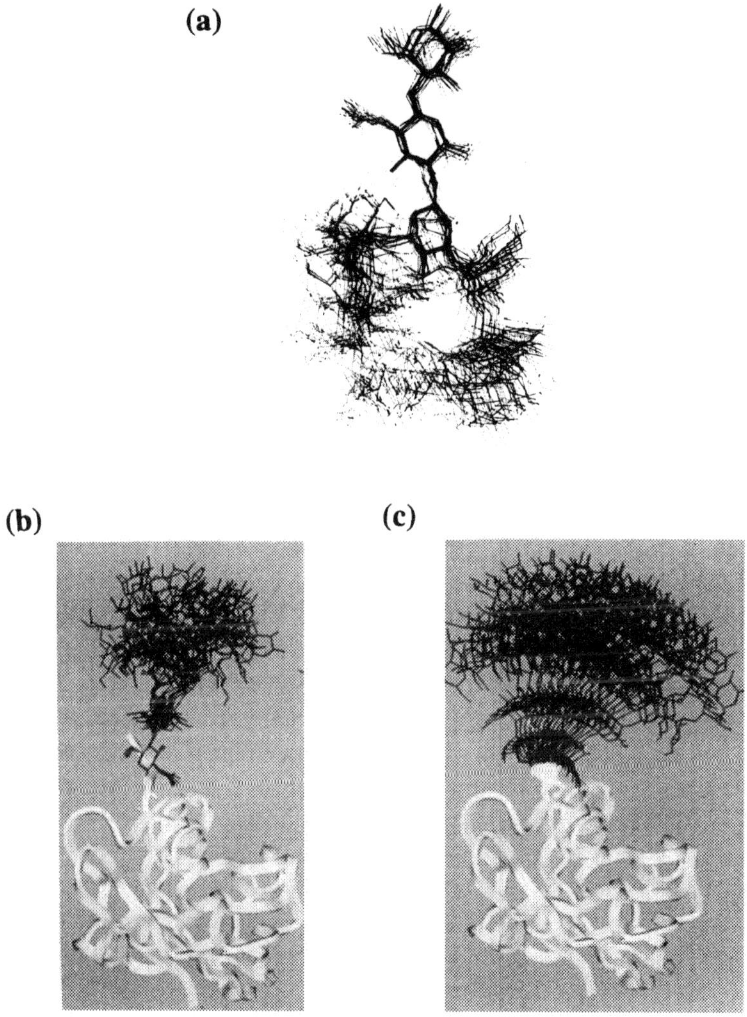

Figure 3. (**a**) A least squares overlay of rings 1–3 from ten snapshots from the trajectory of Man9 each separated in time by 15 ps. (**b**) The Man-9 glycoform of RNase B based on the 2.5 Å X-ray crystal structure with an overlay of 10 oligosaccharide conformations (orange wire frame) from a 750 ps MD trajectory of Man-9 linked through Asn-34. The side chain of Asn-34 was maintained in the crystallographically determined orientation. In order to ensure a correct position for the reducing terminus, the oligosaccharides were overlayed on the first GlcNAc residue. All hydrogen atoms have been omitted for clarity. (**c**) The effect of flexibility of the Asn-34 side chain on the orientation of the oligosaccharide in the Man-9 glucoform of RNase B. The χ_1 and χ_2 angles of the side chain of Asn-34 were varied by ±30° in 15° intervals from the crystallographically determined orientation. The 25 resulting orientations are displayed. All hydrogen atoms have been omitted for clarity.

Dynamic Sugar Model of RNase B

A structural model for RNase B can be constructed from the crystal structure of the protein and the simulation data for Man-9. While the sugar is not resolved in the crystal structure, the side chain of Asn-34 was well defined. In each of the glycoprotein crystal structures discussed above, as well as in those of glycopeptides (Delbaere, 1974; Bush, 1982), the Asn–GlcNAc linkage displays the same conformation. This conformation has been reported also to be present in solution (Bush, 1982; Wormald et al., 1991) (Figure 3b).

Biological Implications

The functional variations associated with glycosylation of RNase have been probed in three ways: (1) by determining the relative abilities of RNase A and B to mediate the hydrolysis of double-stranded RNA (Rudd et al., 1994c), (2) by examining the resistance of RNase A and B to proteases (Rudd et al., 1994c), and (3) by the abilities of antibodies to distinguish between each form. Furthermore, in the case of RNase B, the enzyme activities of several glycoforms have been reported and may be ranked in terms of decreasing activity as: RNase A > RNase Man-0 = RNase Man-1 > RNase Man-5 = RNase B (Rudd et al., 1994c). The enzyme's active site is located in a groove that bisects the protein (Shall and Barnard, 1969). Efficient hydrolysis of RNA necessitates correct alignment of the RNA and the enzyme's active site. This is achieved in part through an interaction between the 5′-terminal phosphate of RNA and a cluster of cationic residues on the protein surface (Lys-31, Lys-37, Arg-10, and Arg-33)(McPherson et al., 1986). Since Asn-34 is present on the surface of the protein near this binding site, it is tempting to speculate that the attenuated RNase activity of the glycoforms, relative to that of the non-glycosylated form, arises from steric hindrance between the oligosaccharide and the RNA. Thus the sugar moiety of a glycoprotein may have a significant effect on the properties of the protein.

Since the conformation of the N-glycosidic linkage is both rigid and planar, the conformational space available to an N-linked oligosaccharide in a glycoprotein may depend to a large extent on the flexibility of the asparagine side chain within the local environment of amino acids (Wormald et al., 1991). However, it is apparent that the molecular volume occupied by the sugar is large in comparison to the single-domain protein and therefore able to shield a large section of the protein surface. When the dynamic nature of the oligosaccharide and the flexibility of the asparagine side chain are also taken into account, the portion of the protein surface covered by the sugar is even more extensive. This indicates that the sugar may interfere with the normal functioning of the protein, including regions of the protein that are considerably removed from the actual linkage site (Figure 3c).

SOME FACTORS WHICH CONTROL PROTEIN GLYCOSYLATION

Some rules have emerged with respect to the factors which control the attachment of oligosaccharides to potential glycosylation sites and the subsequent enzymatic modifications of the glycan chains. While the potential oligosaccharide processing pathways (Kornfeld and Kornfeld, 1985) available to a nascent protein are dictated by the cell in which it is expressed, its final glycosylation pattern is also the result of constraints imposed by the 3D structure of the individual protein.

The Primary Peptide Structure Determines the Number and Location of Potential Glycosylation Sites

The two main classes of glycosidic linkages to proteins (Figure 4) involve either oxygen in the side chain of serine, threonine, or hydroxylysine (O-linked glycans), or nitrogen in the side chain of asparagine (N-linked glycans). To be glycosylated an asparagine residue must form part of the tripeptide AsnXSer where X is any amino acid apart from proline, although the presence of this sequon is not in itself sufficient to ensure glycosylation. The role of the peptide sequence in directing O-glycosylation is less clear, but a Pro-residue, at −1 and +3, may make it favorable. Recently, a consensus sequence (GlyGlySer Thr) has been found to correlate with O-fucosylation in epidermal growth factor domains (Harris et al., 1991; Nishimura et al., 1992; Harris and Spellman, 1993). Other O-linked glycans include those in collagen which are linked through hydroxylysine, and also O-linked GlcNAc residues in the nucleoplasmic and cytoplasmic compartments of cells (Haltiwanger et al., 1992). Physiologically this O–GlcNAc modification is highly labile and seems to be abundant in all eukaryotes (Hart, 1992).

A third type of linkage to proteins has been found for an increasing number of cell surface proteins, which are known to be inserted into the lipid bilayer via a glycophosphatidylinositol (GPI) anchor (Ferguson and Williams, 1988). Only 6 amino acids serve as a GPI attachment site—these are: Cys, Asp, Asn, Gly, Ala, Ser (CDNGAS) (Ferguson, 1991). The amino acids, Gly, Ala, and Ser, predominate at the +1 positions and are obligatory at +2 positions.

Structure and Diversity of N-Linked Glycans

All N-linked glycans contain the pentasaccharide Manα1-6(Manα1-3)Manβ1-4GlcNAcβ1-4GlcNAc as a common core. On the basis of the structure and the location of glycan residues added to the trimannosyl core, N-linked oligosaccharides can be classified into four main groups (Yamashita and Kammerling, 1982; Kornfeld and Kornfeld, 1985). These are: oligomannose, hybrid, complex, and poly-N-acetylgalactosamine (Figure 5).

Oligomannose type glycans (Figure 5-1) contain only α-mannosyl residues attached to the trimannosyl core. Complex type glycans (Figure 5-2) contain no

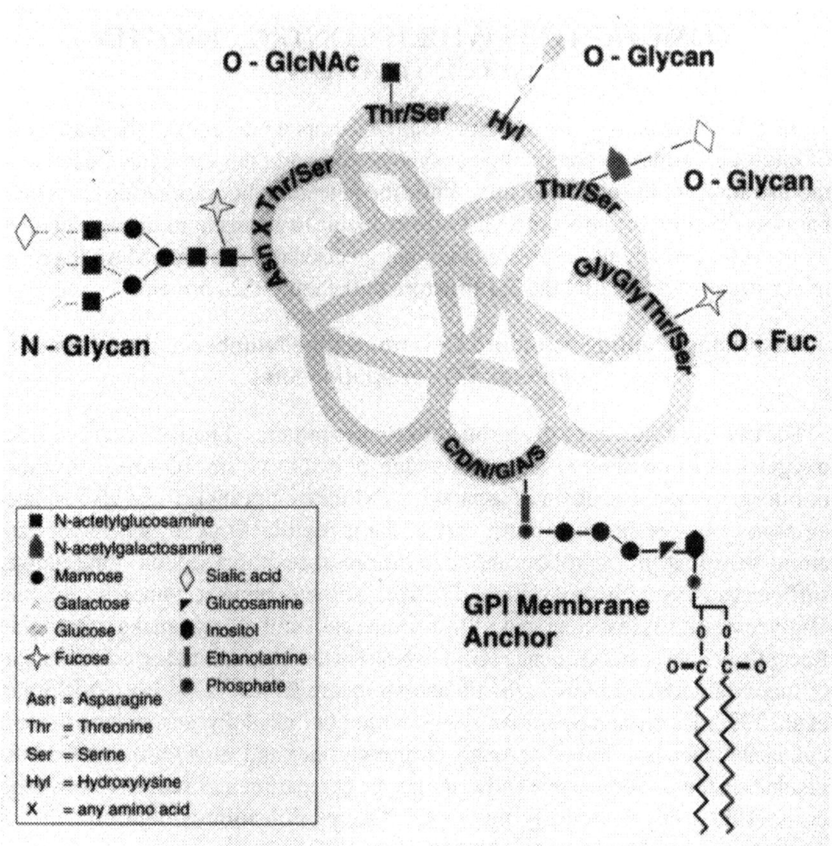

Figure 4. A schematic representation of the main forms of attachment of glycans to a polypeptide. Several glycans may be attached to a single polypeptide and some potential sites may remain unoccupied.

mannose residues other than those in the trimannosyl core, but have "antennae" or branches with *N*-acetylglucosamine residues (Figure 6a) at their reducing termini attached to the core. The number of antennae normally ranges from two (bi-antennary) to four (tetra-antennary), but a penta-antennary structure has been reported in hen ovomuvoid (Yamashita and Kammerling, 1982). While various monosaccharides can be found in the antennae, the presence or absence of fucose and a "bisecting" GlcNAc on the core contributes to the enormous structural variation of complex-type glycans (Figure 6b). Indeed complex-type N-glycans show the largest structural variation in the subgroups resulting mainly from the combinations of different numbers of antennae and variations of monosaccharides in the outer chains. Some of the outer chain structures found in complex type sugar chains are shown in Figure 6b.

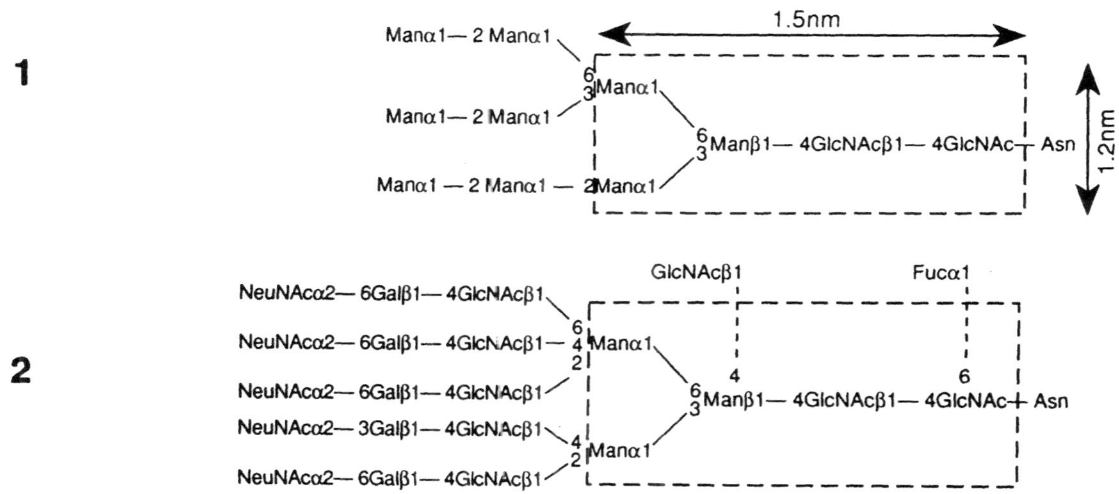

Figure 5. Four groups of N-linked glycans: (**1**) Oligomannose; (**2**) Complex; (**3**) Hybrid type; (**4**) Poly-*N*-acetyl lactosamine(o>m>n). The structure within the shaded box contains the pentasaccharide core common to all N-linked glycans (*continued*).

3

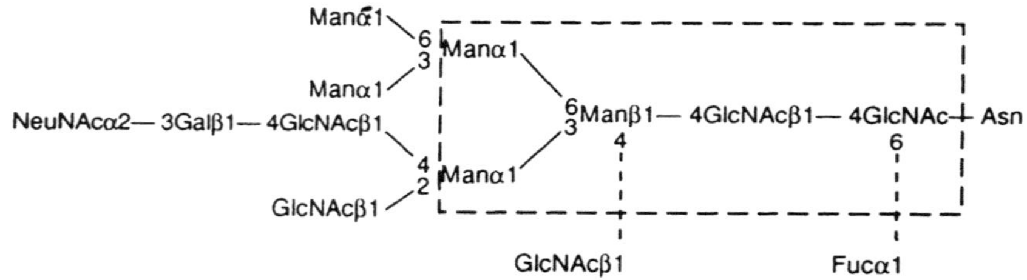

4

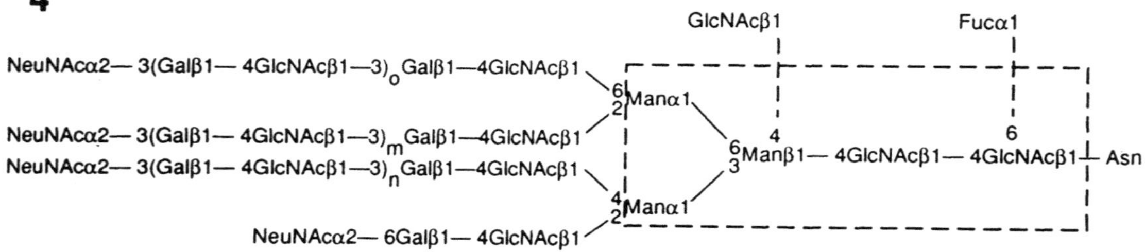

Figure 5. *(continued)*

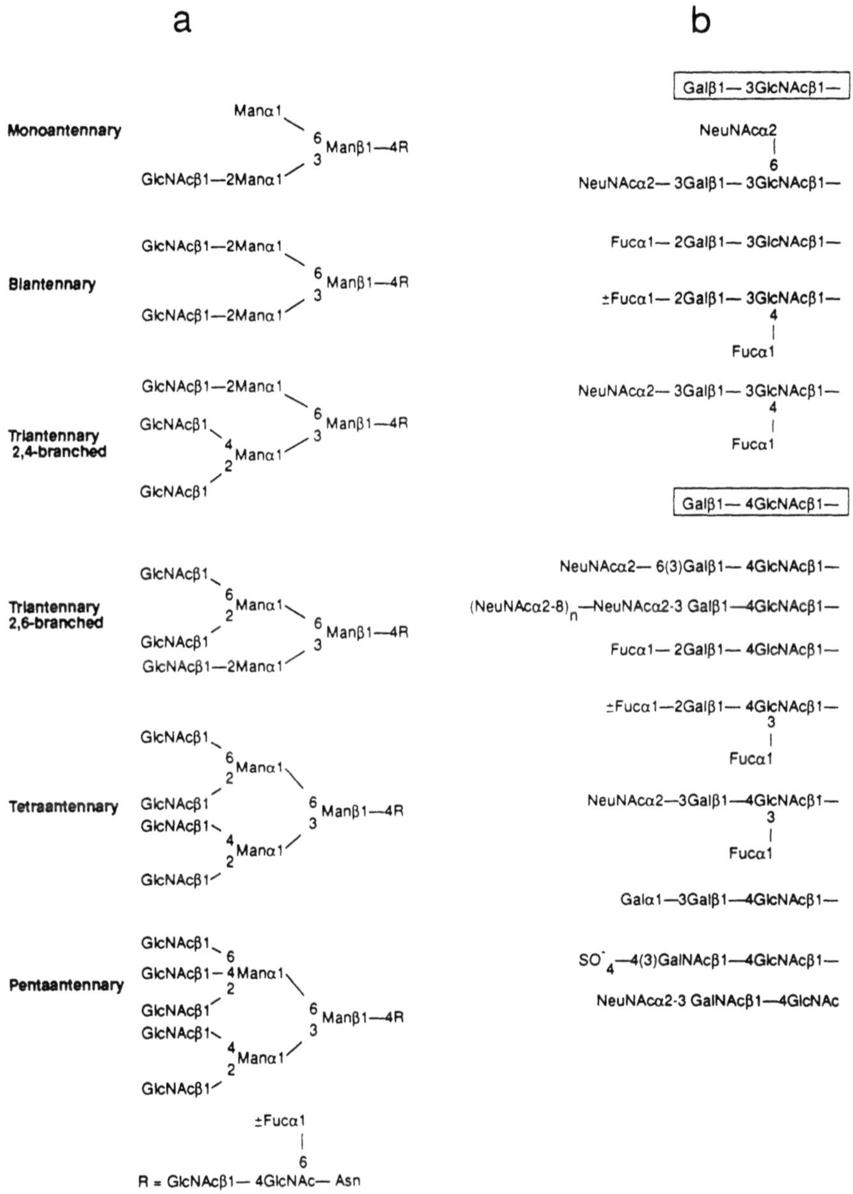

Figure 6. Two major elements that create the diversity of structures of complex type sugar chains: (**a**) branching differential and (**b**) variations in chain structures.

The hybrid type N-glycans (Figure 5-3) have the characteristic features of both complex-type and high mannose-type glycans. One or two α-mannosyl residues are linked to the Manα1-6 arm of the trimannosyl core (as in the case of oligomannose type glycans) and usually one or two antennae (as found in complex type glycans) are linked to the Manα1-3 arm of the core.

The fourth group (Figure 5-4) is the poly-N-acetylgalactosamine N-glycans containing repeating units of (Galβ1-4GlcNAcβ1-3-) attached to the core. These repeats are not necessarily uniformly distributed on the different antennae and the galactosamine repeat may also be branched. Poly-N-acetylgalactosamine extensions are most frequently found in tetra-antennary glycans (Fukuda, 1994).

Structure of O-Linked Glycans

In contrast to N-linked glycans, O-linked glycans do not share a common core structure. They are based on a number of different cores (Schachter and Brockhausen, 1992). So far they can be categorized into at least six groups according to different core structures (Figure 7). These cores can be elongated to form the backbone region by addition of Gal in β1-3 and β1-4 linkages, and GlcNAc in β1-3 and β1-6 linkages. Although the glycans are often linked to serine or threonine residues through GalNAc, the linkages may be through other residues, e.g. fucose. We should also note that single glycans, such as fucose or GlcNAc, may be O-linked to the peptide backbone. O–GlcNAc can also be β-linked as found in cytoplasmic and nucleoplasmic proteins (Haltiwanger et al., 1992).

Cell Type Influences Glycosylation

The type of cell has a major role in determining the extent and type of glycosylation, which is both species- and tissue-specific (Parekh et al., 1989a). Oligosaccharides are formed on an "assembly line". For protein-bound and lipid-bound oligosaccharides this is the endoplasmic reticulum (ER) and the Golgi apparatus (Kornfeld and Kornfeld, 1985). A series of membrane-bound glycosidases and glycosyltransferases act sequentially on the growing oligosaccharide as it moves through the lumen of the ER and Golgi apparatus. Many different enzyme reactions (typically eight for a complex oligosaccharide such as those on IgG) are involved in the processing pathways. Each individual enzyme reaction may not go to completion, giving rise to *glycoforms* or glycosylated variants of the polypeptide. The type of enzymes (glycosidases and transferases), their type, concentrations, kinetic characteristics, and compartmentalization, reflect both the external and internal environment of the individual cell in which the protein is glycosylated. This explains why the glycosylation patterns of natural glycoproteins may be influenced by physiological changes such as pregnancy, and also by some diseases which may affect one or more of the enzymes in the cell. For example, in IgG isolated from rheumatoid arthritis patients the galactosyl transferase activity may be decreased. This results in an alteration of the glycoform populations if the Fc is

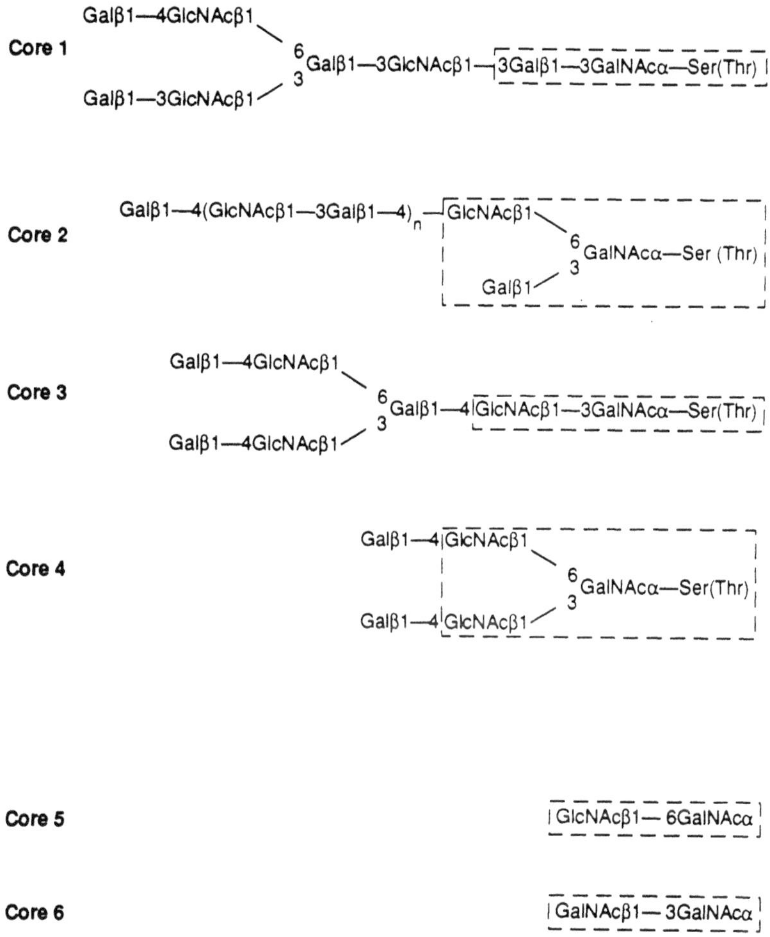

Figure 7. Six types of core structures (boxed) among those found in O-linked glycans.

altered, reflecting an increase in the proportion of agalactosyl N-linked glycans. Glycosylation is often an exquisite indicator of the "health" of a cell. However, the factors which control the expression of the enzymes in the assembly line remain to be elucidated.

The glycosylation of recombinant glycoproteins can be very sensitive to changes in conditions, such as the glucose concentration of the culture medium (Goochee and Monica, 1990). The glycosylation pattern basically reflects the type of cell used in the expression system and the use of different cell lines can result in significant glycosylation differences. For example, there are differences in the branching structure in the complex-type oligosaccharides. These arise from the different

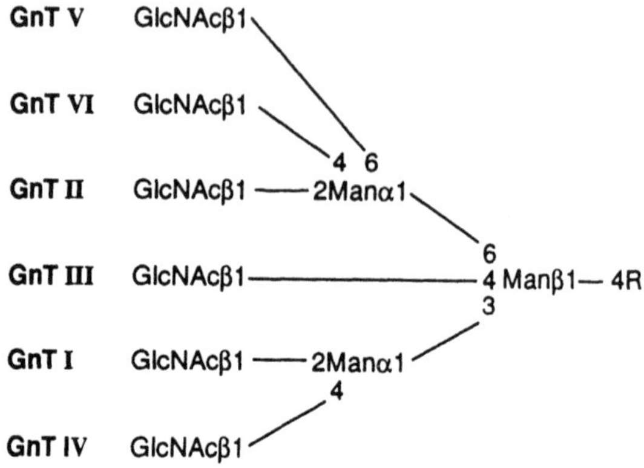

Figure 8. The "branching" GlcNAc-transferases. Five antennae can be initiated on the (Manα1-3)Manβ1-4GlcNAcβ1-4GlcNAcβ-Asn core of *N*-glycans by the actions of GlcNAc-transferases I, II, IV, V, and VI. A "bisecting" GlcNAc can be added by GlcNAc-transferase III.

expression of the GlcNAc transferases shown in Figure 8 (Schachter, 1986; Schachter, 1994).

The 3D Structure of the Protein Influences the Extent and Type of Glycosylation

The 3D structure of the individual protein clearly has a role in determining the type and extent of its glycosylation. A number of mechanisms may be involved. These include:

1. The position of the glycosylation site in the protein. N-linked sites at the exposed turns of β-pleated sheets, which are sometimes close to proline residues, are normally occupied while those near the C-terminus are more often vacant.
2. Access to the glycosylation site on the developing oligosaccharide. This may be sterically hindered by the local protein structure or by protein folding which may compete with the initiation of N-glycosylation.
3. Interaction of the developing oligosaccharide with the protein surface. This may result in a glycan conformation which may alter the accessibility to specific glycosyltransferases or glycosidases.
4. Interaction of the glycosyl enzymes with the protein structure. This can lead to site-specific processing.
5. Glycosylation at one site in a multiglycosylated protein. This may sterically hinder events at a second site on the same molecule.

6. The interaction of protein subunits to form oligomers. This may prevent glycosylation or restrict the glycoforms at individual sites.

Although the same glycosylation machinery is available to all the proteins which are translated in a particular cell and use the secretory pathway, it has been estimated that between 10 and 30% of potential glycosylation sites are not occupied (Mononen and Karjalainen, 1987; Gavel and Von Heijne, 1990). Moreover site analysis has shown that the distribution of different classes of N-linked oligosaccharide structures is frequently specific for each site on a protein. In the case of rat brain Thy1, for example, site 23 contains only oligomannose structures, site 74 has only complex and hybrid, while all three classes of glycans are present at site 98 (Parekh et al., 1987; Williams et al., 1993).

Generation of Glycoforms

The initial event in N-linked glycosylation is the cotranslational transfer, to an asparagine residue within a glycosylation sequon, of the dolichol-linked $Glc_3Man_9GlcNAc_2$ oligosaccharide to the nascent polypeptide chain (Figure 9). A series of trimming events then occurs. First, glucosidase I hydrolyzes the outermost glucose (α1-2) residue, followed by the removal of the remaining two α1,3 glucose

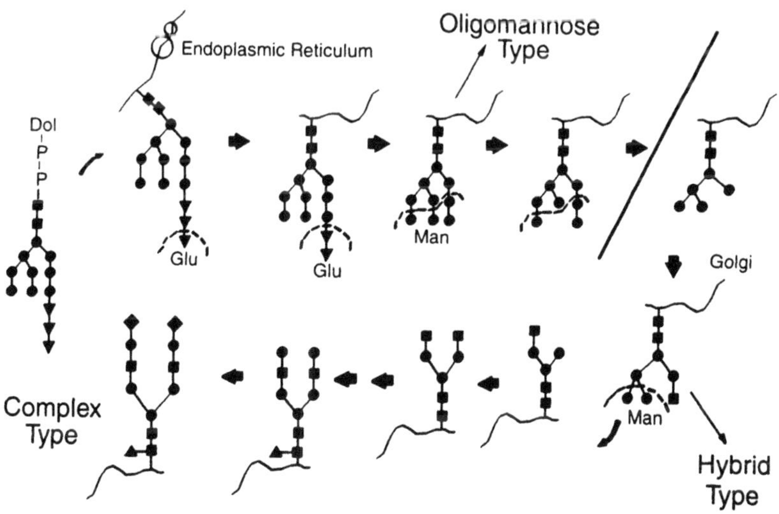

Figure 9. Representation of some of the steps in oligosaccharide biosynthesis (Kornfeld and Kornfeld, 1985). The symbols represents the following monosaccharides: ▾ glucose, ● mannose, ■ NAcetylglucosamine, ♦ sialic acid, ▴ fucose.

residues by glucosidase II. These reactions are reasonably rapid (~minutes) and the protein is assumed to be fully folded by this stage. Subsequent enzyme reactions may clearly be influenced by the 3D structure of the protein in respect of the accessibility of the individual enzymes. The routing of glycoproteins within the cell, the compartmentalization of trimming enzymes with different specificities and the competing secretion pathways are also important factors controlling the biosynthesis of N-linked oligosaccharides.

The synthesis of O-linked oligosaccharides is entirely a posttranslational event with a series of enzymes acting sequentially on the fully folded protein. Many of the factors discussed above will also apply. Initially O-linked oligosaccharides are covalently attached through an O-glycosidic monosaccharide, and a serine or threonine. Some of the enzymes that act subsequently may be found in both the N- and O-linked biosynthetic pathways.

CHARACTERISTICS OF PROTEIN GLYCOSYLATION

There are three levels of understanding of protein glycosylation. Members of the immunoglobulin superfamily—CD4 and CD2 (Ashford et al., 1993; Davis et al., 1993), and Thy-1 (Parekh et al., 1987; Williams et al., 1993) (Figure 10)—well illustrate the main points.

Importance of the Overall Protein Conformation in Determining Glycosylation

The chromatographic gel filtration profiles of the sugars released from soluble recombinant forms of human CD4, rat CD4, and rat CD2 expressed in Chinese hamster ovary (CHO) cells (Davis et al., 1990, 1993) are shown in Figure 11. The glycosylation potential in CHO cells has been well characterized (see references in Dwek et al., 1993a) and yields multi-antennary and poly-*N*-acetylgalactosamine oligosaccharides. Rat soluble CD2 (sCD2) showed glycosylation typical of the CHO cell line with bi-, tri-, and tetra-antennary complexes, and with hybrid structures and the poly-*N*-acetylgalactosamine species (Figures 11 and 12) (Davis et al., 1993). In contrast, despite the available repertoire of processing enzymes, the N-linked glycans of rat and human soluble CD4 (sCD4) expressed in CHO cells had quite different glycosylation profiles. Most of the oligosaccharides were of the bi-antennary complex, hybrid or oligomannose type (Carr et al., 1989; Harris et al., 1990; Spellman et al., 1991; Ashford et al., 1993). These results indicate the importance of protein structure in determining the pattern of glycosylation.

Effect of Local Protein Conformation–Glycosylation Shows Site Specificity

The extent of processing CD4 is less than in CD2. As CD2 is structurally very similar to the first two domains of CD4 (Jones et al., 1992) (Figure 10) the three-dimensional conformation of these members of the immunoglobulin superfamily cannot be the only factor influencing their glycosylation. The local amino acid sequence and microenvironment of the glycosylation site must also be important.

To illustrate this, the site specificity of glycosylation in wild-type rat CD4 was determined by isolating glycopeptides containing the glycosylation sites at Asn-270 and Asn-159 (Ashford et al., 1993). The glycosylation patterns at each site were different (Figure 13). In particular, oligomannose and hybrid structures were restricted to Asn-159, the nonconserved site. The conserved site (Figure 10) contained exclusively biantennary complex oligosaccharides as had been found for the equivalent site in human sCD4 (Spellman et al., 1991). Therefore overall differences in glycosylation between rat and human glycoproteins can be accounted for by site-specific glycosylation at the nonconserved sites. Detailed analysis showed that there were three oligosaccharide structures associated with Asn-270 and 10 with Asn-159 giving an ensemble for CD4 of 30 glycoforms.

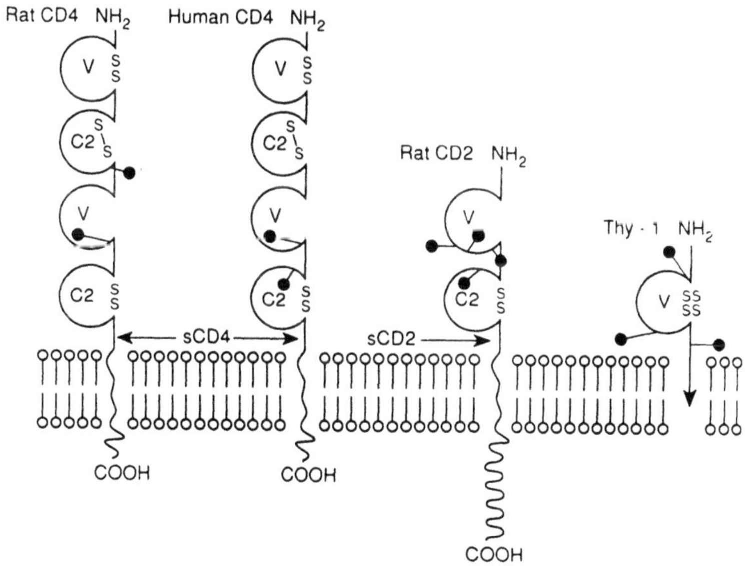

Figure 10. Schematic drawings of rat and human CD4, rat CD2, and Thy-1. The molecules are drawn with the circles representing immunoglobulin superfamily (IgSF) domains and the "lollipops" N-linked oligosaccharides. The glycosylphosphatidylinositol membrane anchor of Thy-1 is depicted as a vertical arrow. The IgSF domains are designated as V or C2 on the basis of sequence analysis (Williams et al., 1989). The positions of the mutations introduced in the CD4 and CD2 molecules to produce the recombinant soluble forms are indicated by horizontal arrows (adapted from Williams et al., 1989).

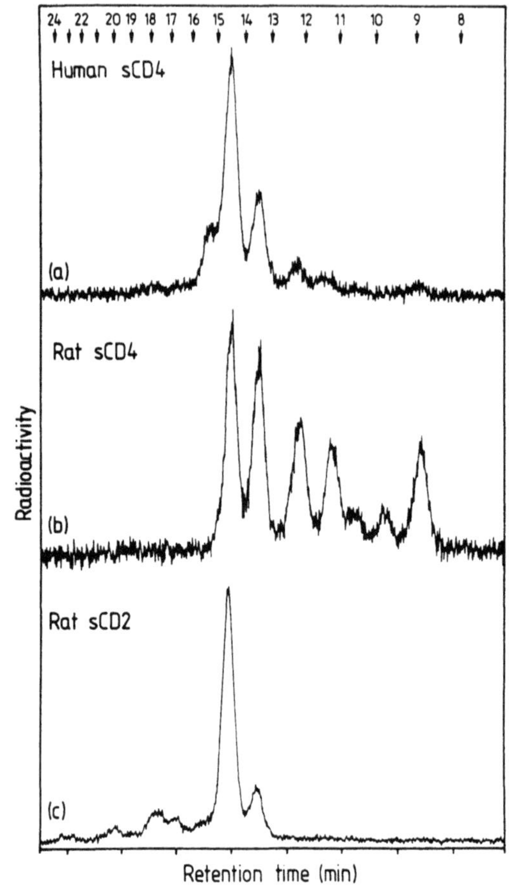

Figure 11. Bio-Gel P-4 gel filtration profiles of the desialylated, tritium-labelled oligosaccharides of recombinant soluble CD4 and CD2 expressed in CHO cells. (**a**) Total oligosaccharides of human sCD4. (**b**) Total oligosaccharides of rat sCD4. (**c**) Total oligosaccharides of rat sCD2. The vertical arrows indicate the elution positions of isomalto-oligosaccharides containing the corresponding number of glucose units. The time axis is marked at 100 min intervals (data taken from Ashford et al., 1993; Davis et al., 1993).

A further question arises whether the processing at each site is independent of the processing at the other. One approach is to produce mutants with the appropriate glycosylation sites deleted. In this case the glycosylation patterns from the variants with either Asn-270 or Asn-159 mutated show strong similarity to that from the glycosylated peptides of the wild-type (Figure 13). It can therefore be concluded that specific and independent processing occurs at each glycosylation site.

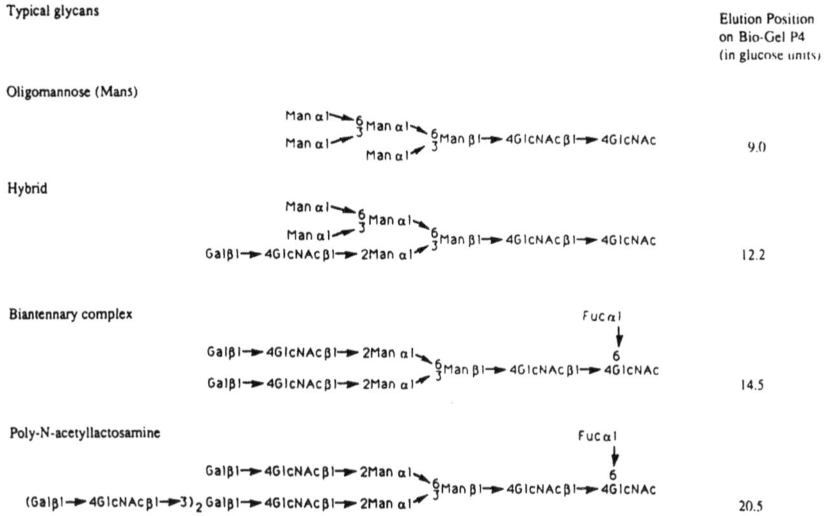

Figure 12. Examples of the structure of the different types of neutral desialylated N-linked oligosaccharides. Fuc, L-fucose; Gal, D-galactose; GlcNAc, D-N-acetylgluc-samine; Man, D-mannose.

Glycosylation is Protein-Specific, Site-Specific, and Tissue/Cell-Specific Thy-1

The characteristics of N-glycosylation at individual sites have been probed within a single immunoglobulin domain, thereby eliminating inter-domain effects. Thy-1, a GPI membrane-anchored molecule, has one (V-type) immunoglobulin-like domain and three N-glycosylation sites (Figure 10) (Williams and Gagnon, 1982). A comparative study of rat Thy-1 from thymocytes and brain showed tissue specificity of N-glycosylation (Parekh et al., 1987), although their amino acid sequences are identical. The differential effects of tissue glycosylation were expressed at the level of individual glycosylation sites. The site distribution of oligosaccharides was such that no Thy-1 molecules were found in common between the two tissues (Figure 14). Thus each tissue created unique sets of glycoforms; a glycoprotein must be viewed as an ensemble or collection of glycoforms.

Tissue Plasminogen Activator, tPA

tPA is a glycoprotein protease consisting of five domains: a fibronectin "finger" domain, an epidermal growth factor (EGF) domain, two kringles, and the catalytic serine protease domain (Ny et al., 1984; Patthy, 1985). Binding sites for fibrin are present within the finger and kringle domains. There are two main classes of

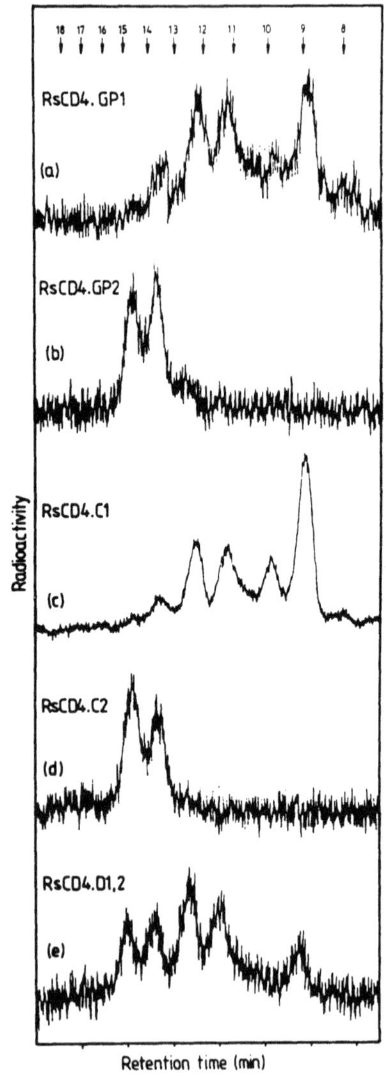

Figure 13. Bio-Gel P-4 gel filtration profiles of the desialylated, tritium-labelled oligosaccharides of the glycopeptides and glycosylation mutants of rat soluble CD4. (**a**) Total oligosaccharides of rat sCD4-derived glycopeptide from the region of the first glycosylation site. (**b**) Total oligosaccharides of rat sCD4-derived glycopeptide from the region of the second glycosylation site. (**c**) Total oligosaccharides of rat sCD4 with the second glycosylation site removed. (**d**) Total oligosaccharides of rat sCD4 with the first glycosylation site removed. (**e**) Total oligosaccharides of rat sCD4. NH_2-terminal-two IgSF domain form. For details of the annotation see legend to Figure 2 (reproduced from Figure 8 Ashford et al., 1993).

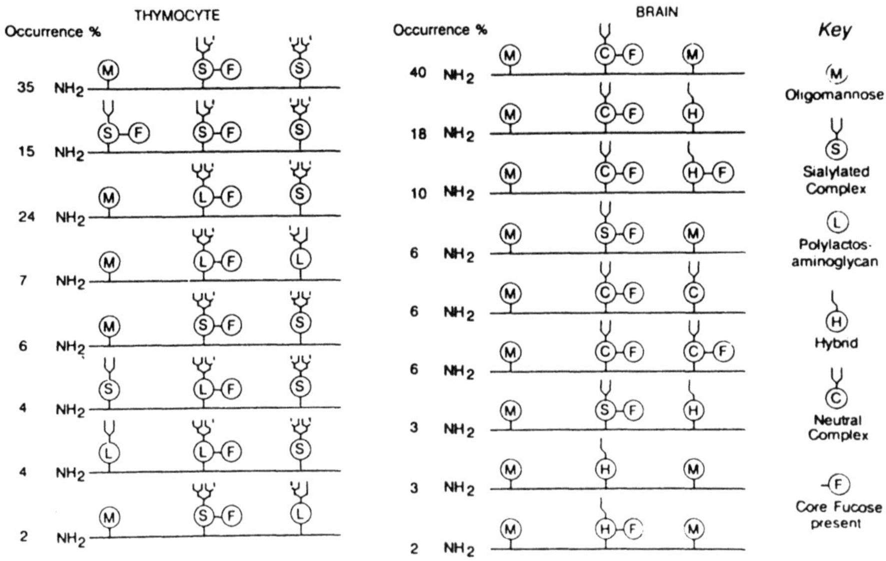

Figure 14. A comparison of the nature and percentage composition of the glycoforms of thymocyte and brain-derived rat Thy-1 glycoprotein. Glycoforms representing less than 2% abundance are not illustrated.

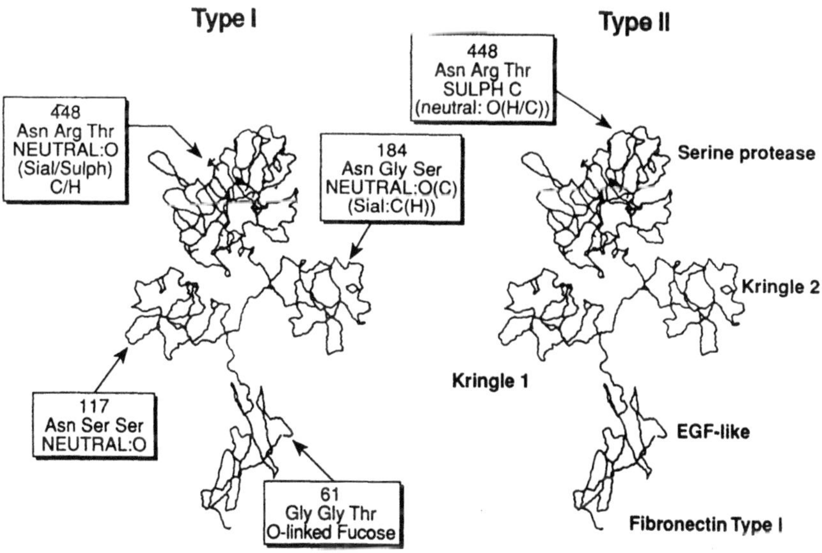

Figure 15. Site specificity of glycosylation in tPA. Note that the classes of glycan structures present at Asn 448 on tPA depend on the site occupancy of Asn 184, while those at site 117 do not.

glycosylated variants of the molecule, type I and type II. Type I has three N-linked sugars at Asn-117, Asn-184, and Asn-448 (Figure 15), whereas type II has only two at Asn-117 and Asn-448. Each of these classes has a subpopulation of glycoforms. Variable occupancy of site 184 affects the fine structure of the glycan population at site 448, demonstrating that glycosylation at one site can influence the processing at another. Human colon fibroblast tPA has a different set of non-overlapping glycoforms from melanoma tPA, where in Bowes melanoma type I tPA the major species at site 448 were found to be neutral glycans of the complex or hybrid structures. In type II, however, 72% of the structures were sulfated complex glycans (Figure 15) and there were relatively few neutral sugars (Parekh et al., 1989). This suggests that in tPA the presence of a glycan at site 184 shields the developing glycan chain at site 448 from some of the glycosylation processing enzymes available in the cell.

GLYCOSYLATION MODULATES ENZYME ACTIVITIES

The effects of glycosylation on the properties of ribonuclease were discussed earlier. Other examples of modulation are shown by tPA and plasminogen, which in contrast to RNase with its single domain, are multidomain proteins.

tPA

Various properties of type I tPA are affected by the occupancy of site 184:

1. Glycosylation at site 184 hinders the plasmin-mediated conversion of single to two-chain tPA (Wittwer and Howard, 1990).
2. The fibrinolytic activity of type II tPA on plasminogen exceeds that of type I tPA regardless of the cell line in which the tPA is produced. Variable occupancy of site 184 is one of the factors which controls the rate at which plasmin is generated, and this may be related to the differences in the affinity of type I and type II tPA for lysine (Wittwer et al., 1989).

Plasminogen

Plasminogen, the natural substrate of tPA, is a multidomain protein consisting of five kringle regions and a serine protease domain. Plasminogen is a mixture of two major glycoforms (Brockway and Castellino, 1972; Spellman et al., 1989) which have the same amino acid sequence (Sottrup-Jensen et al., 1978) and contain one O-glycosylation site and a potential N-glycosylation site in kringle 3. Both type I and type II plasminogen contain an O-linked sugar chain at Thr-345, while the N-glycosylation site at Asn-288 contains a biantennary sugar in type I, but is unoccupied in type II (Hayes and Castellino, 1979). Human recombinant plasminogen, expressed in *E. coli* and consequently not glycosylated, is resistant to activation by tPA (Gonzales-Gronow et al., 1990).

The importance of the O-linked glycoforms in plasminogen 2 has been demonstrated by Pirie-Shepherd and co-workers (1995) who isolated six glycoforms differing in sialic acid content. The kinetic activity of the different glycoforms was decreased as the sialylation increased.

Various properties of plasminogen are also altered by the presence of the N-linked oligosaccharide (type I) perhaps due to significant shielding of the kringle 3 domain of the protein by the N-linked glycan.

Some of the properties which are altered include:

1. A slower rate of the β- to α-conformational change from an open to a compact form (Ponting et al., 1992).
2. A 10-fold weaker binding of plasminogen to U937 cells (Gonzalez-Gronow et al., 1989).
3. A lower affinity for lysine sepharose (Hayes and Castellino, 1979).
4. A slower rate of activation of plasminogen by urokinase.

SOME STRUCTURAL ROLES FOR OLIGOSACCHARIDES

Glycosyl–Phosphatidylinositol (GPI) Anchors

Since 1985 over 100 examples of GPI anchored proteins have been described (Ferguson, 1991). The GPI anchor is an alternative anchoring mechanism to the transmembrane polypeptide domain of type I membrane proteins. The first detailed structural studies on GPI anchors, of *T. brucei* variant surface glycoprotein (VSG) and of rat brain Thy-1, were performed in 1988 (Ferguson et al., 1988; Homans et al., 1989). In each case, the C-terminus of the protein is linked by ethanolamine phosphate to a glycan with the conserved backbone sequence Manα1-2Manα1-6Manα1-4$GlcNH_2$, which in turn is linked to the sixth position of the *myo*-inositol ring of phosphatidylinositol.

The conserved sugar sequence may be a consequence of the method of biosynthesis in which a preassembled core is added to a protein in much the same way as N-glycosylation results in general in a common core. In both cases, variations on this may be protein- and cell-specific. In VSG this tetra-saccharide backbone is substituted with branched side chains of α-galactose. The arrangement of the GPI anchor of the VSG protein may permit the close packing of these molecules on the surface of the parasite, with the heterogeneity of the glycan part of the GPI anchor allowing close packing in two-dimensional space in order to create a surface. The structure in solution (Homans et al., 1989) showed that the glycan could exist in an extended configuration along the plane of the membrane spanning an area of 600 $Å^2$, which is similar to the cross-sectional area of the monomeric N-terminal VSG domain (Figure 16).

By contrast, computer modeling of the structure of Thy-1 and its GPI anchor suggests that the lipid part of the GPI may not be the sole membrane anchor

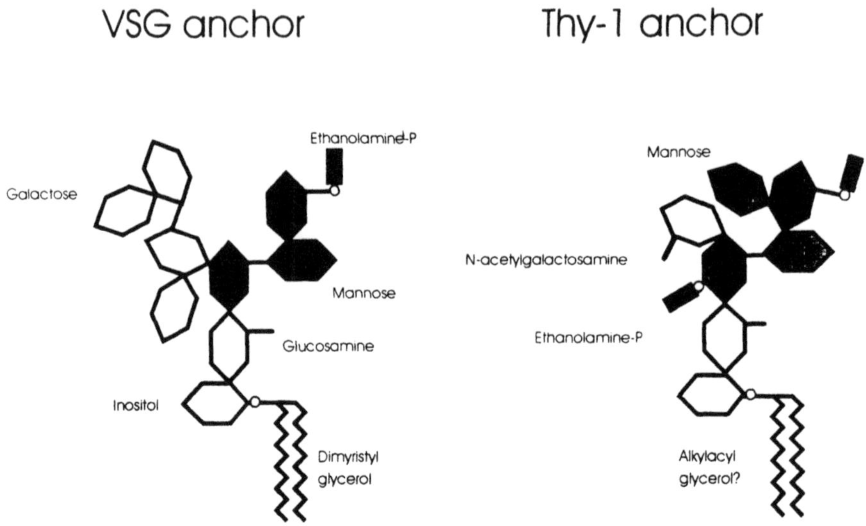

Figure 16. Comparison of the structures of the VSG and Thy-1 glycan anchors.

(Rademacher et al., 1991). The glycan part of the GPI may lie within the Thy-1 protein moiety so that the Thy-1 protein moiety sits directly on the membrane with most of the GPI actually within the protein. This smaller anchor would allow access to Thy-1 by other molecules within the membrane.

Structure–Function Relationships in IgG

X-ray crystallography has shown that each region of homology in the IgG molecule corresponds to a compact, independently folded unit, and that these are linked together by short sections of polypeptide chain. Each domain consists of two β-pleated sheets with antiparallel strands connected by loop regions (the immunoglobulin fold) (Amzel and Poljak, 1979). Crystallographic studies of IgG Fc fragments (Deisenhofer et al., 1981; Sutton and Phillips, 1983) have shown that, unlike other immunoglobulin domains, the two C_H2 domains do not form extensive lateral associations. The resulting interstitial region accommodates the complex oligosaccharides, which are attached to Asn-297 on each heavy chain. For rabbit Fc, unlike that of humans, one of the α1-3 arms interacts with the trimannose core of the opposing oligosaccharide. In human Fc, the α1-3 arm may interact with the protein but there is a better definition for the α1-6 arm of each oligosaccharide. These interact with the hydrophobic and polar residues Phe-243 (Man-5 and GlcNAc-6), Pro-244/245 (Gal-7) and Thr-260 (GlcNAc-6 and Gal-7) on the domain surface. The recognition depends on the presence of subsites for sugars on the protein involving both the aromatic residues and hydrogen bond formation between the protein and water molecules.

The effects of glycosylation may be subtle and highly specific. Both the C1q and protein A binding sites on IgG are located in the C_H2 domain between Phe-319 and Ile-332 (distal or at the carboxyl-terminal side from the N-linked glycosylation site). Neither of these functions is markedly affected by the absence of carbohydrate (Leatherbarrow and Dwek, 1983; Leatherbarrow et al., 1985). The binding of monocytes, which involves sites 234–237 (proximal to the amino-terminus) on the lower hinge region, is eliminated in aglycosylated monoclonal murine IgG. This is consistent with a proposal that the absence of sugars results in a lateral movement of domains in the hinge region relative to the normally glycosylated antibody. Moreover, in aglycosylated IgG a protein structural change has been detected by ^1H NMR at His-268 (Lund et al., 1990), which is also in the vicinity of the lower hinge. These data suggest that the oligosaccharide may stabilize a particular hinge conformation essential for monocyte binding. The spatial relationship between the C_H2 and C_H3 domains, on the other hand, does not seem to depend on the presence of the oligosaccharides because in their absence protein A and C1q binding are unaffected.

OLIGOSACCHARIDE RECOGNITION

Specific Interactions with Animal Lectins

In order to decode the information present in oligosaccharides they must be recognized by other molecules—lectins. Following oligosaccharide recognition, lectins mediate many different biological processes. These include immune defense (e.g. the mannose binding protein and the macrophage mannose receptor), clearance of glycoproteins (e.g. the asialo–glycoprotein receptor), and cell–cell adhesion (e.g. the selectins). This important aspect of the glycobiology derives mainly from the work of Drickamer (1988). Approaching 100 lectins have been isolated and classified principally on the basis of sequence homology. By far the largest class is the C-type lectins. These bind carbohydrates in a Ca^{2+}-dependent manner. The lectins contain *carbohydrate-recognition domains* (CRDs). These CRDs contain a common sequence motif of approximately 120 amino acid residues and are characterized by 31 invariant or highly conserved amino acids (Weis et al., 1991). Some of the C-type CRDs which have been reported are shown in Figure 17.

One C-type lectin is the mannose-binding protein (MBP) (see below) which mediates antibody-independent binding of pathogens containing a high concentration of mannose or *N*-acetylglucosamine residues on their surface. The recognition event can lead to complement fixation or opsonization (Ikeda et al., 1987). Recent X-ray data on the CRD from this protein illustrates the principles involved in the recognition of oligosaccharides. The CRD shows specificity for terminal residues and involves the sugar chelating to the Ca^{2+} ion via the 3- and 4-hydroxyl groups

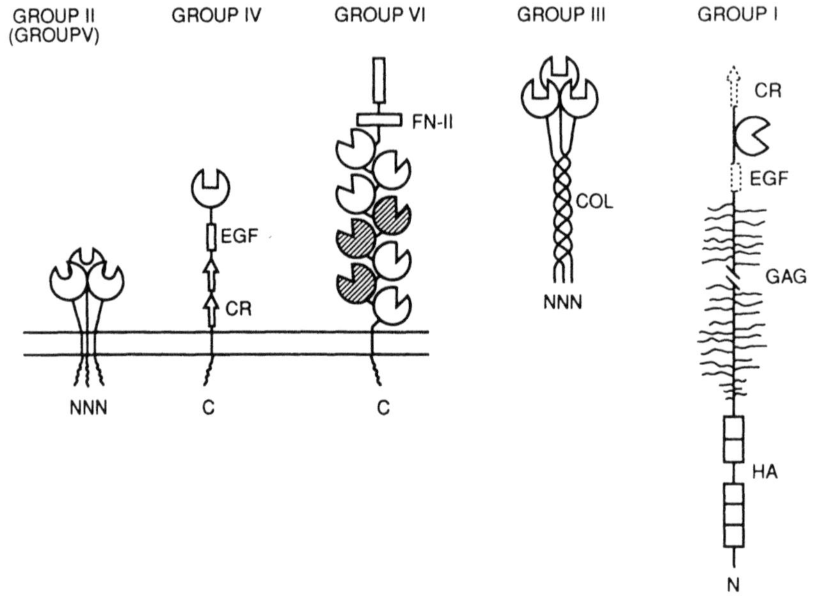

Figure 17. Summary of the structures of several groups of C-type animal lectins. Representative structures for three groups of membrane-associated lectins are shown: Group II, the chicken hepatic lectin (homologue of the mammalian asialoglycoprotein receptor); Group IV, the selectin cell adhesion molecules; and Group VI, the macrophage mannose receptor. Group III lectins (collectins), such as mannose-binding protein, are found in extracellular fluids. Group I CRD-containing proteins are proteoglycans of the extracellular matrix. Other domains present in these molecules include EGF, epidermal growth factor-like repeats; CR, complement regulatory domains; FN-II, fibronectin type II repeats; COL, collagen-like sequences, GAG, glycosamino-glycan attachment sites; and HA, hyaluronic acid-binding domains (reprinted from Weis et al., 1992).

(Weis et al., 1992a,b). There are also interactions between the CRD and the monosaccharide residues in the oligosaccharide that are mediated by water molecules, effectively increasing the surface area of the oligosaccharide in contact with the CRD. C-type lectins and their constituent CRDs display weak affinity for monosaccharides.

Triggering biological events often requires multivalent receptors interacting with multivalent ligands. Multivalency in the receptor can be achieved by clustering CRDs. In the MBP, for example, this occurs by the formation of oligomers of polypeptide chains, each of which contains a single CRD. In contrast the macrophage mannose receptor contains CRDs in a single polypeptide. Clustering of CRDs could also arise from having multiple copies of the lectin in close proximity. (This may be the case for the selectins.) Similarly, the oligosaccharide may be multivalent by virtue of its branching if it has the "correct" geometry, or again there may be multiple copies appropriately presented on a surface.

The recent X-ray structures of the MBP trimer (Sherrif et al., 1994; Weis and Drickamer, 1994) can be used to illustrate the above points. The CRDs in the trimers are separated by some 44–53 Å. Matching multivalency by an oligomannose oligosaccharide to the trimer array can only be obtained if the terminal residues have the correct geometry in that they must also be separated by similar spacings of 44–53 Å. NMR studies with molecular dynamic calculations have shown that in Man_9, which has three terminal residues, the maximum distances between them is approximately 21 Å (Woods et al., 1994a). Consequently the Man_9 structure can be bound via only one terminal residue to MBP, but this monovalent binding will not result in the biological triggering of complement. (This would only result from multiple presentation of Man_9 structures—as on a pathogen—with the appropriate spacings between the individual oligosaccharides to bind to the different CRDs.) In this way the MBP molecule can "inspect" a range of molecules in the serum. Those with a single oligomannose structure will be bound and released—while these multiply presented as on a pathogen will lead to the biological effector functions being triggered. This type of mechanism has all the hallmarks of the discrimination in the immune system between self and nonself.

Another receptor, which is believed to form part of a basic defense mechanism against pathogens, is the macrophage mannose receptor (Drickamer and Taylor, 1993; Mullin et al., 1994; Taylor, 1993). This receptor can also mediate the elimination of endogenous proteins such as tissue plasminogen activator (via the oligomannose structures at site Asn-117) and lysosomal enzymes which carry high mannose-type oligosaccharides. Such proteins are often released into the circulation in response to pathological events. The macrophage mannose receptor internalizes the bound ligands and targets these to lysosomes for destruction.

Neural Glycosylation and Recognition

Glycosylation is a common feature of molecules implicated in cell–cell and cell–matrix interactions in the development and maintenance of the nervous system—as is involved, for example, in axon growth, guidance and targeting. There is now increasing evidence to suggest that oligosaccharides play key recognition roles in these processes. A knowledge of oligosaccharide structures involved is essential for an understanding of carbohydrate mediated interactions at the molecular level.

One way to probe for functional effects of glycosylation is to test glycans from the brain libraries as specific competitive inhibitors in neurite outgrowth assays *in vitro*. Wing et al. (unpublished) have shown that crude sugar fractions from a rat brain library (at a concentration of 500 μM) can inhibit neurite outgrowth over astrocytes, a process involving recognition between cell adhesion molecules.

Carbohydrates are enriched at synapses and, as functions for the associated glycoproteins are becoming established, the exciting possibility exists for a glycoconjugate role in synaptic efficacy, a field with far-reaching implications in such areas as memory formation.

Major Histocompatibility Complex (MHC) Restricted Recognition of Glycopeptides by T-Cells

The major histocompatibility complex (MHC) class 1 molecules display peptides from self or foreign cellular proteins, on the antigen-presenting cell surface. CD8+ cytolytic T-cells probe these peptide MHC complexes to identify and subsequently eliminate cells that express foreign peptides.

An interesting and exciting consequence of the specific recognition of carbohydrates has been the demonstration that T-cells recognize glycopeptides bound to MHC Class I molecules. Haurum et al. (1994) have reported the efficient binding to Class I MHC of a synthetic glycopeptide carrying the naturally occurring O-β-linked *N*-acetylglucosamine (GlcNAc). The glycosylated molecule is highly immunogenic, and elicits a specific, MHC-restricted, anti-glycopeptide cytotoxic T-cell response. Recognition of glycosylated peptides could therefore be of great importance in the acquisition of immunity towards malignant diseases and viral infections.

Recognition of Oligosaccharides by Stimulated T-Cells

A quite different type of T-cell recognition involving oligosaccharides may exist in diseases such as rheumatoid arthritis, Behcet's syndrome, and IgA nephropathy. In these conditions up to 9% of the T-cells are bound by cytophilic IgA1 compared with none in resting T-cells from healthy subjects. Detailed studies have shown that the binding to this subset of T-cells can be inhibited by the "O"-linked sugars associated with the hinge region of IgA1 (Rudd et al., 1994a), suggesting that oligosaccharide recognition may have functional implications in these diseases.

Although the general area of T-cell recognition of saccharides is a difficult and challenging one, it is an important area which is well worth pursuing and which could provide yet more insights into the roles of oligosaccharides in the cellular immune response.

GLYCOSYLATION IN DISEASE

The IgG Molecule and Rheumatoid Arthritis

The IgG molecule contains on average 2.5 oligosaccharide chains per molecule. Two of these represent the conserved glycosylation sites in the Fc portion of all IgGs at Asn-297. The remainder occur in the hypervariable regions of the Fab fragment with a frequency and position dependent on the occurrence of an N-glycosylation site (Asn/Xaa/Ser(Thr)). The glycosylation is of the complex biantennary class and about 30 variants occur in the IgG molecule (Figure 18). This glycosylation of IgG results in many different glycoforms.

There are clear site-specific differences in glycosylation between the Fab and Fc fragments with regard to sialylation and the presence of bisecting GlcNAc residues (both mainly in Fab). Characteristics of Fc N-glycosylation include a low incidence (10%) of monosialyted structures, the absence of disialylated structures, a low incidence of cores carrying a "bisecting" GlcNAc, and heterogeneity in the galactose residues. In general, the Fc oligosaccharides are mainly restricted to bianten-

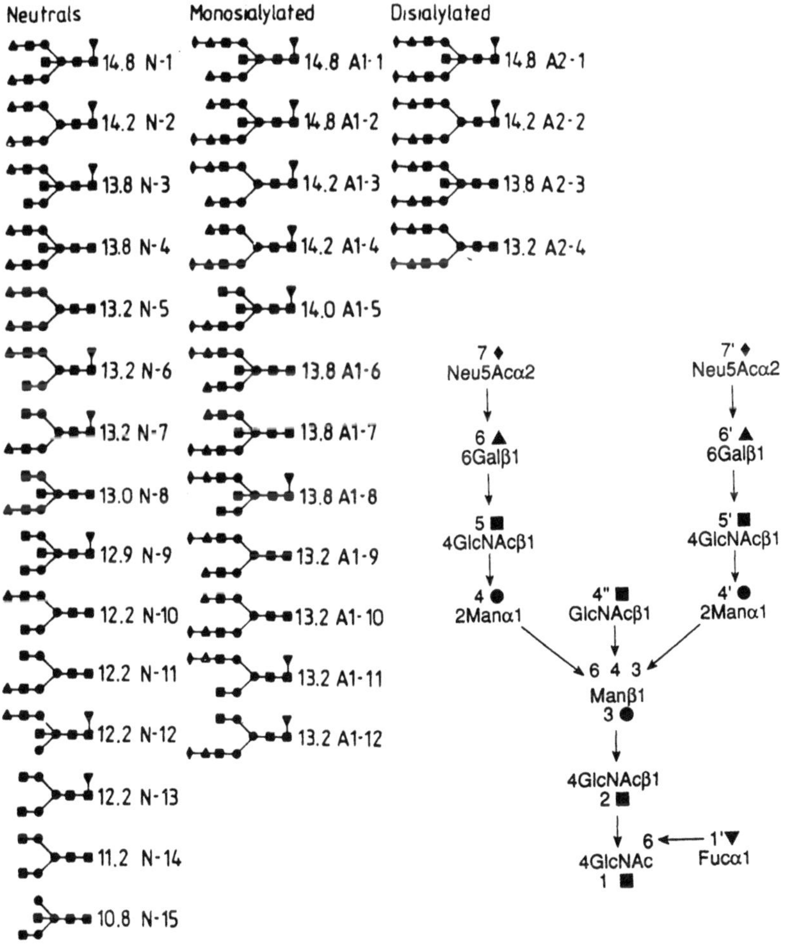

Figure 18. Primary sequences of the N-linked oligosaccharides associated with IgG. The hydrodynamic volume of each structure, measured in glucose units, is indicated.

nary oligosaccharides terminating in 2, 1, and 0 galactose residues (G2, G1, and G0). The unique 3D environment in the Fc may limit the accessibility of sugar-processing enzymes so resulting in these glycoforms. Fab N-glycosylation is characterized by a high incidence of di- and monosialylated structures, and of cores with the "bisecting" GlcNAc residue (Rademacher et al., 1986). It should be stressed that the large number of different structures associated with IgG is not the result of studying a polyclonal population, since a similar heterogeneity is found upon analysis of myeloma and hybridoma-derived IgG.

Serum IgG from patients with rheumatoid arthritis contains the same set of bi-antennary oligosaccharides found in normal individuals, but in very different proportions (Figure 19). The incidence of structures with outer-arm galactose is dramatically decreased, and the incidence of those structures terminating in outer-arm *N*-acetylglucosamine correspondingly increased (Parekh et al., 1985). A comparison of the N-glycosylation of Fab and Fc fragments derived from total serum IgG of patients with rheumatoid arthritis, or from a control group, shows that the decreased galactosylation is largely due to changes in the N-linked oligosaccharides of the Fc. There are also quantitatively minor, but potentially significant differences

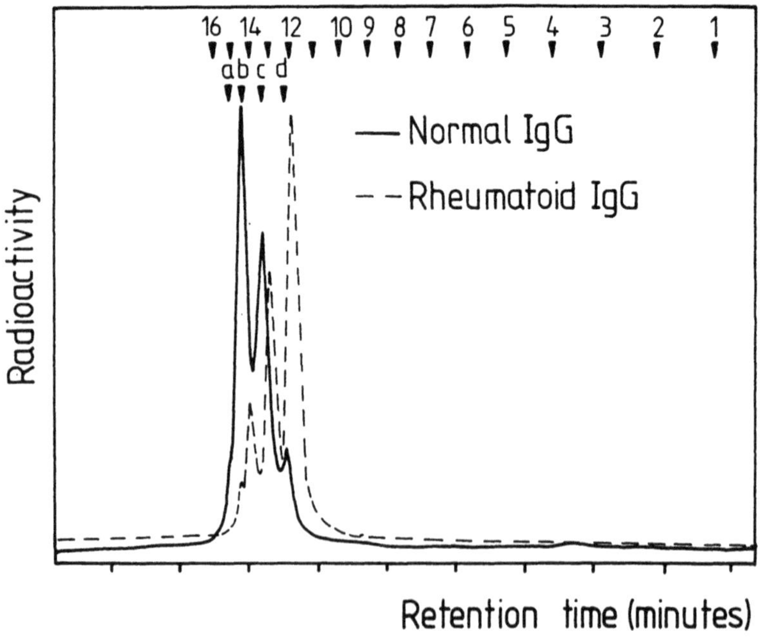

Figure 19. Representative Bio-Gel new P-4 (-400 mesh) Gel permeation chromatogram of the asialo oligosaccharides of total serum IgG from a healthy individual and a patient with rheumatoid arthritis.

in Fab glycosylation (Scragg and Chang, unpublished). The changes in glycoforms occur in all four IgG subclasses (Youings and Dwek, unpublished).

The change in galactosylation of the serum IgG of patients with rheumatoid arthritis is not common to all autoimmune or inflammatory disorders. Agalactosyl IgG has been consistently found in patients with juvenile rheumatoid arthritis, Crohn's disease, and tuberculosis (Parekh et al., 1988, 1989b).

Fc glycoform distribution varies with age, with the severity of rheumatoid arthritis, and with pregnancy (Figure 20). These changes may reflect the control of galactosyl transferase activity under different physiological conditions. Rheumatoid arthritis was shown to be associated with changes in G0 glycoforms. This parameter is therefore a good biochemical marker of diagnostic value in disease (Figure 20). In an arthritic woman with pathologically elevated levels of G0 glycoforms, changes in G0 correlated with remission of arthritis during gestation and postpartum recurrence (Rook et al., 1991).

That the G0 glycoforms may be an important factor in rheumatoid arthritis can be shown experimentally from the arthritis induced in mice by collagen (CIA)

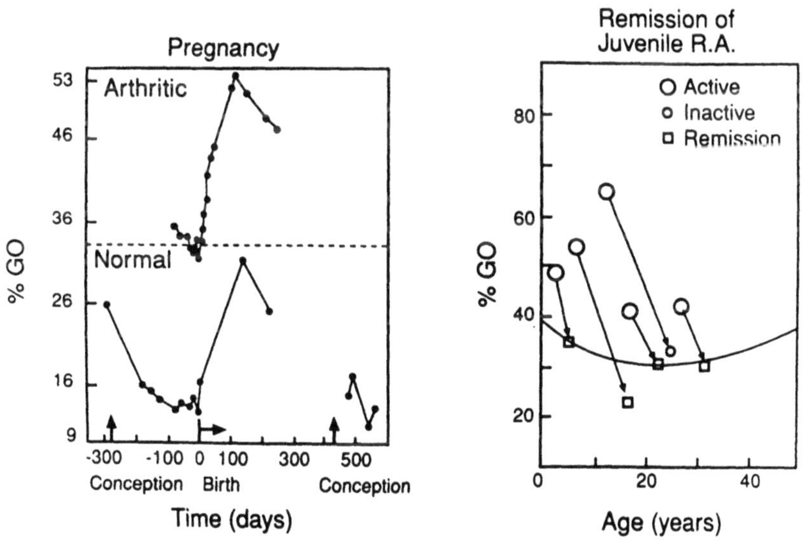

Figure 20. Variations in the galactosylation of IgG. There are changes with age, and disease activity as illustrated for juvenile arthritis. A comparison between the changes in percentage G0 during the course of pregnancy for a normal and rheumatoid arthritis patient, shows that both patients tend to increase their galactose levels during pregnancy. The dashed line indicates the aged matched expected value of a normal healthy individual. It is seen that the pregnant patient with rheumatoid arthritis achieves this level and this correlates with a remission of arthritis in the patient.

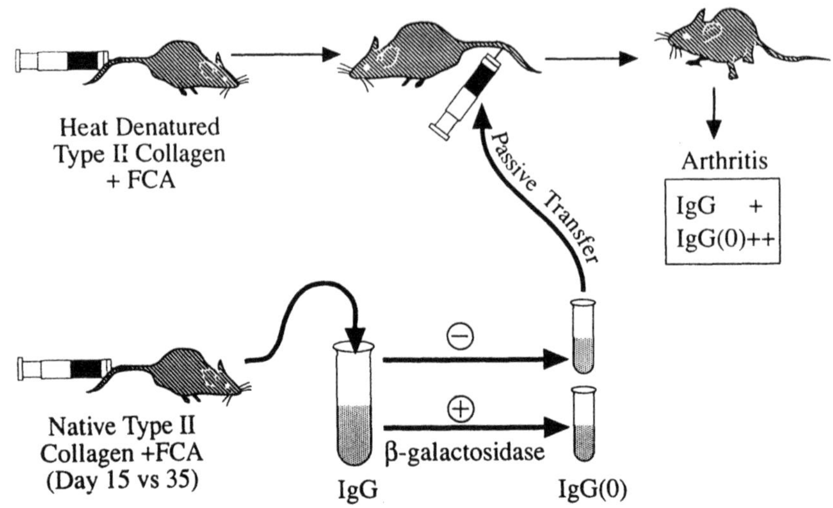

Figure 21. Protocol for passive transfer of IgG in the collagen induced arthritis model in mouse. Purified IgG is transferred to a suitably primed mouse, or is first treated with β-galactosidase to enrich the IgG0 glycoforms.

(Figure 21). IgG was purified from pooled sera of mice with CIA at days 17 and 38 when peak levels of type II collagen autoantibodies were present. The IgG fractions were isolated and divided. One fraction was treated with β-galactosidase from *Streptococcus* strain 6646K to generate IgG(G0) glycoforms exclusively. By increasing the level of agalactosyl IgG glycoform of the anti-type II collagen antibodies present in the preparation, the IgG became more effective in causing arthritis. This indicates that the glycosylation status of an autoantibody is one factor in determining if an antibody is pathogenic.

Structural Implications of IgG(G0)

Galactose–Amino Acid Interaction

Studies of the Fc fragment by X-ray crystallography and NMR indicate that galactose residues present on both antennae can interact with amino acid residues on the protein. An increase in the level of IgG glycoforms lacking terminal galactose and thereby terminating in *N*-acetylglucosamine could lead to the exposure of certain Fc determinants. This may elicit an immune response, or raise a preexisting subclinical response to a pathological one which may be relevant to rheumatoid arthritis. In addition, the now vacant galactose sites on the protein may create a lectin-like activity in the IgG resulting in the formation of complexes or autoaggregates typical of the disease (Roitt et al., 1988) without an actual autoimmune response.

The Fc region of antibody molecules mediates interaction with many of the effector functions of the immune system following antigen binding. Of these, the complement system is a major immune defense mechanism. Inappropriate or chronic localized complement activation can cause severe damage to host tissue, and this is an important factor in the pathogenesis of several diseases (Morgan, 1990). The first step in the classical complement cascade is the binding of C1q to the C_H2 domains in the Fc region of the antibody, following antigen recognition. It involves surface "matching" of charged amino acid residues between C1q and Fc (Burton et al., 1980; Duncan and Winter, 1988). Activation of the classical complement pathway by a second route, which does not involve C1q, is mediated by the serum lectin, mannose binding protein (MBP), and until now has been reported as antibody-independent (Ikeda et al., 1987; Malhotra et al., 1994). Serum MBP recognizes pathogens which have a high concentration of mannose or GlcNAc residues on their surface. Using NMR and model building in conjunction with X-ray data, it has been demonstrated that the terminal sugar (GlcNAc) becomes exposed and accessible to MBP only in those molecules in which the Fc oligosaccharides lack galactose.

Interactions still cannot occur (due to protein–protein steric interactions) without displacing the oligosaccharide from the position observed in the X-ray structure (Deisenhofer et al., 1981). NMR data show that such displacements occur spontaneously on loss of galactose, and molecular modeling indicates that these make either the 3-arm or 6-arm terminal GlcNAc residues available for binding to the carbohydrate recognition domain.

Ca^{2+}-Dependent Binding of MBP to IgG is Mediated by the Agalactosyl Fc Glycoforms

There is an increase in specific MBP binding when normal IgG (G0 = 20%) is converted enzymatically to 100% IgG(G0) (Figure 22). Comparison of the binding data for IgG(G0) and the corresponding Fab and Fc fragments with those of normal IgG and its fragments indicates that the increased binding arises from interactions between MBP and the Fc-associated oligosaccharides. Although normal IgG and its Fab and Fc fragments also contain G0 oligosaccharide structures, the data suggest that their density and presentation to the multivalent MBP is insufficient to give rise to strong binding. In IgG it is, therefore, the alteration in levels of the Fc-glycoforms containing G0 structures that modulate MBP binding to IgG. Figure 22b shows a representative IgG preparation from a rheumatoid patient (G0 = 36%) which exhibits increased binding of MBP.

MBP Activation of Complement by Agalactosyl IgG Glycoforms

Figure 23 shows the activation of the complement system by MBP following binding to IgG or IgG(G0) immobilized on microtitre plates. The amount of C4 activated and deposited was dependent on the concentration of MBP (Figure 23a).

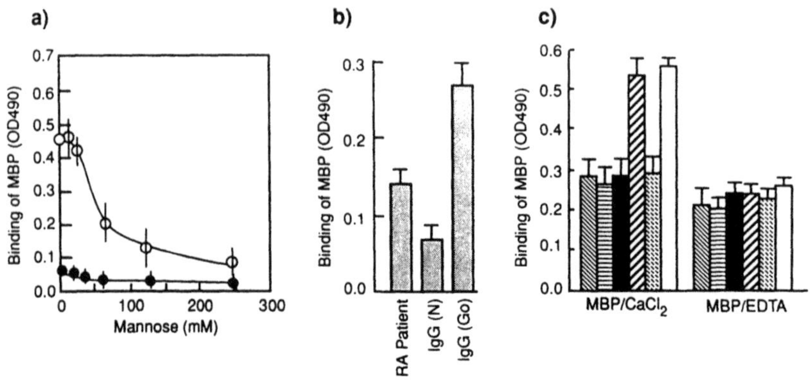

Figure 22. Interaction of MBP with IgG, IgG-G0 and their fragments. (**a**) The Ca^{2+} dependent binding of MBP to IgG-G0 (O) or IgG (●) is inhibited in a concentration-dependent manner by mannose. The binding of MBP in the presence of EDTA (O.D. value of 0.3) was subtracted from the data to take account of nonspecific binding. The error bars represent the range of three different determinations. All experiments were done in triplicate. (**b**) The binding of MBP in the presence of Ca^{2+} to normal IgG (20% G0), IgG from an RA patient (36% G0), and IgG-G0 (100% G0). The background value of O.D. = 0.06 was subtracted from the data to take account of nonspecific binding (in EDTA). (**c**) Glycoforms of normal IgG and Fc containing only G0 type sugars show a higher Ca^{2+} dependent binding of MBP than the unmodified normal populations of IgG and Fc. Fc (▨), Fab (▤), IgG ●, Fc-G0 (▨), Fab-G0 (▨), IgG-G0 (□).

At the saturation point the amount of C4 fixed was ca. fivefold higher when MBP was complexed with IgG(G0) compared with normal pooled IgG. MBP-mediated activation of complement is induced predominantly by the interaction between MBP and the Fc/G0 and not the Fab/G0 fragment (Figure 23b). Fixation of C4 was reduced considerably when MBP was incubated with IgG, IgG(G0), or their fragments in the presence of mannose, an inhibitor of binding of MBP to oligosaccharides (Figure 23c). This demonstrates that the activation by MBP is mediated through the IgG(G0) sugars. In contrast, the activation by C1q is independent of the glycosylation state of the IgG (Figure 23d) confirming that treatment with glycosidases has not altered the conformation of the IgG.

MBP and Agalactosyl IgG are Present in Synovial Fluid

Levels of IgG(G0) are elevated in the synovium compared with serum (Rademacher et al., 1988b; Tsuchiya et al., 1993). MBP in synovial fluid is very similar to that in serum (Malhotra et al., 1995). Its presence, coupled with the high levels of IgG(G0), suggests that activation of complement by MBP could contribute to the chronic inflammation of the synovial membrane of affected joints. This would provide a link between the elevated levels of G0 glycoforms found in rheumatoid arthritis and the pathogenesis of the disease.

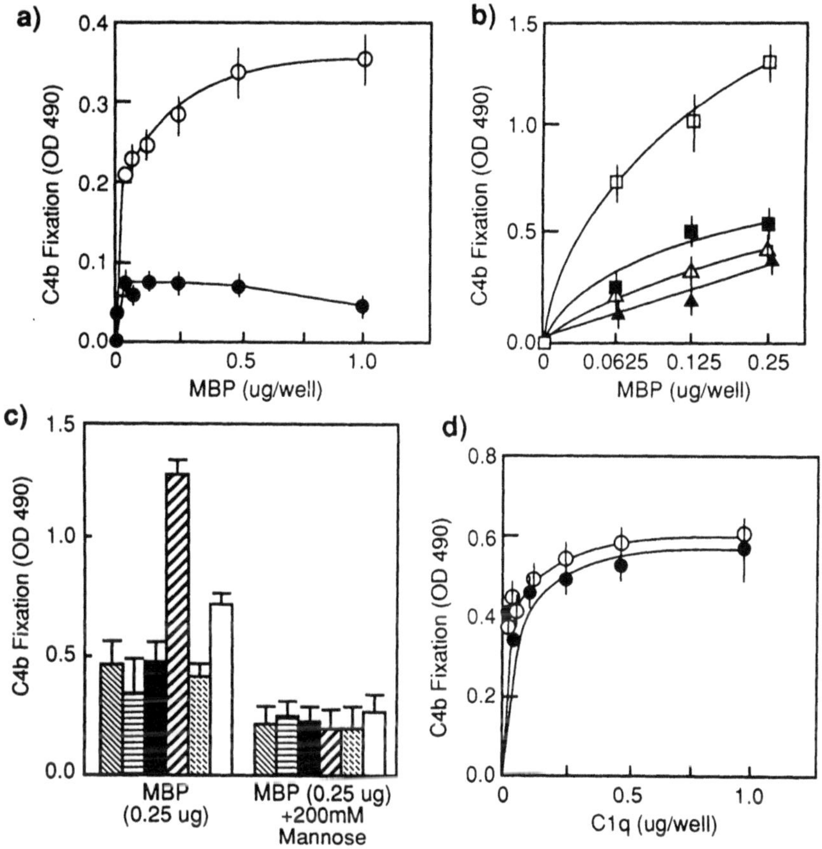

Figure 23. IgG-G0-induced activation of the complement system. (**a**) IgG-G0 (O) induces 5 times more deposition of C4b than IgG (●). Deposition of C4b is dependent on MBP concentration. The nonspecific binding was reflected in an O.D. of 0.18 which was subtracted from the data points. (**b**) Increase in C4b deposition occurs when Fc (■) is converted to Fc/G0 (□). There is no change in C4b deposition when Fab (▲) is converted to Fab/G0 (△). The background value of O.D. = 0.19 was subtracted from each data point. (**c**) Binding of MBP is through the oligosaccharide. A complement activation assay was performed as in Figure 5 a, b, except that MBP was incubated in the protein coated wells in TBS/Ca^{2+} or TBS/Ca^{2+} containing mannose. The incubation of MBP with Fc ( ) or Fab ( ) or IgG (■) or Fc/G0 ( ) or Fab/G0 ( ) or IgG-G0 ( ) in the presence of mannose decreased the deposition of C4b. A background value of O.D. = 0.19 was subtracted from each data point. (**d**) Conversion of IgG (●) to IgG-G0 (○) does not markedly alter the C1q-mediated deposition of C4b. Serial dilutions of C1q (100 µl; 10 µg/ml) were incubated in IgG-G0 or IgG coated wells and C1q-mediated deposition of C4b was measured as above. The background value of O.D. = 0.18 was subtracted from each data point in the IgG curve, while a value of 0.32 was subtracted from the IgG-G0 points to account for background due to the residual MBP present in the depleted serum.

INHIBITORS OF GLYCOSYLATION AS ANTIVIRAL AGENTS

Studies with HIV

Two glucosidases are involved in the biosynthesis of N-linked oligosaccharides (Grinna and Robbins, 1979): Glucosidase I, which removes the terminal α1,2-linked glucose residue; and glucosidase II hydrolyzing the remaining two α1,3-linked glucose residues. Subsequent processing to complex and hybrid type structures takes place through the action of mannosidases and glycosyl transferases (Hubbard and Ivatt, 1981; Kornfeld and Kornfeld, 1985).

Several inhibitors of purified glucosidases have been identified (Datema et al., 1987; Winchester and Fleet, 1992) which would be expected to block complex type oligosaccharide synthesis. However, in some systems when cells are treated with these compounds, complex type oligosaccharide formation still occurs. There are several possible reasons for this. These include not achieving a high enough inhibitor concentration within the endoplasmic reticulum, and the presence of endomannosidase activity which would provide a bypass mechanism to circumvent glucosidase inhibition (see Lubas and Spiro, 1987). Also the effects of glucosidase inhibition on cellular glycoproteins are selective. Some glycoproteins require correct oligosaccharide processing for secretion or cell surface expression, while for others complete processing of their oligosaccharides is less critical.

The envelope glycoproteins of HIV are heavily N-glycosylated. HIV-1 gp120 has 20–25 potential sites for N-linked glycosylation, with the carbohydrate contributing 50% of its apparent molecular weight (Lasky et al., 1986). The positions of the glycosylation sites within the primary amino acid sequence of gp120 are relatively consistent between different isolates of HIV-1 (Alizon et al., 1986; Willey et al., 1986; Leonard et al., 1990) (Figure 24). At least 13 of the glycosylation sites are conserved, and the remaining sites usually are not located more than approximately 10 residues from the sites in the reference strain, HIV-1_{IIIB} (Leonard et al., 1990). Studies have shown that a diverse range of high mannose, hybrid, and bi-, tri-, and tetra-antennary structures are present both on recombinant and virally derived gp120, and that the proportions of the different glycoforms are similar between the two systems (Geyer et al., 1988; Mizuochi et al., 1988a,b, 1990). Analysis of gp120 mutants suggests that N-glycosylation of either gp120 or gp41 is necessary for post-CD4 binding events, such as the fusion of the viral and cellular membranes (Willey et al., 1988; Lee et al., 1992).

When a range of sugar analogues were screened for anti-HIV activity *in vitro* (Fleet et al., 1988; Karpas et al., 1988), *N*-butyldeoxynojirimycin (NB-DNJ) was found to be a potent inhibitor of infection and exhibited minimal cytotoxicity. This compound inhibited purified α-glucosidase I with a K_i of 0.22 μm. NB-DNJ also inhibited glycoprotein processing in intact cells at concentrations equivalent to those which inhibited HIV replication *in vitro* (Karlsson et al., 1993). Treatment with antiviral concentrations of NB-DNJ results in the terminal sequence

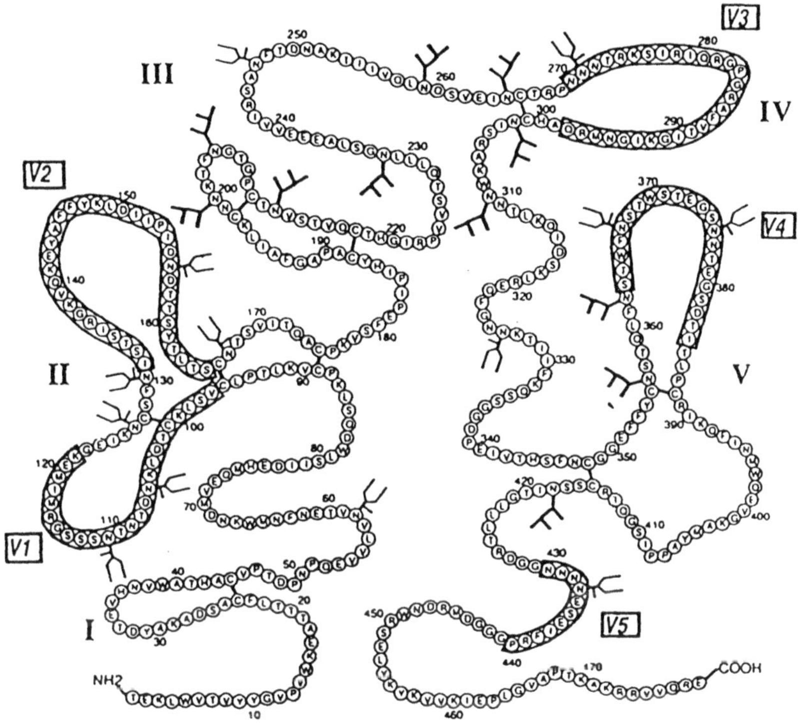

Figure 24. Schematic diagram of the HIV-1 envelope glycoprotein gp120, adapted (Leonard et al., 1990). Glycosylation sites containing high mannose and/or hybrid oligosaccharides and those containing complex type oligosaccharides are indicated. The disulphide-bonded domains are labelled with roman numbers and the hypervariable regions are enclosed in boxes and labeled V1–V5.

Glcα1,2Glcα,1,3Glcα1,3Man being present on gp120 oligosaccharides (Figure 25).

This has also been demonstrated for gp120 derived from H9 cells, acutely infected with the HIV-1 IIIb strain (Karlsson, 1993). Two consequences of treatment with NB-DNJ are the inhibition of synctia formation in cells infected with HIV-1, and a reduction in release of infectious virus. This reduction is not caused by a decrease in the release of virus particles but from a reduced infectivity of the virus (Figure 26).

Glycosyl Transferase Inhibitors and Glycosphingolipid Storage Diseases

NB-DNJ is a glucose analogue and, in addition to inhibiting glucosidases, it inhibits the glucosyltransferase-catalyzed biosynthesis of glucosylceramides. This

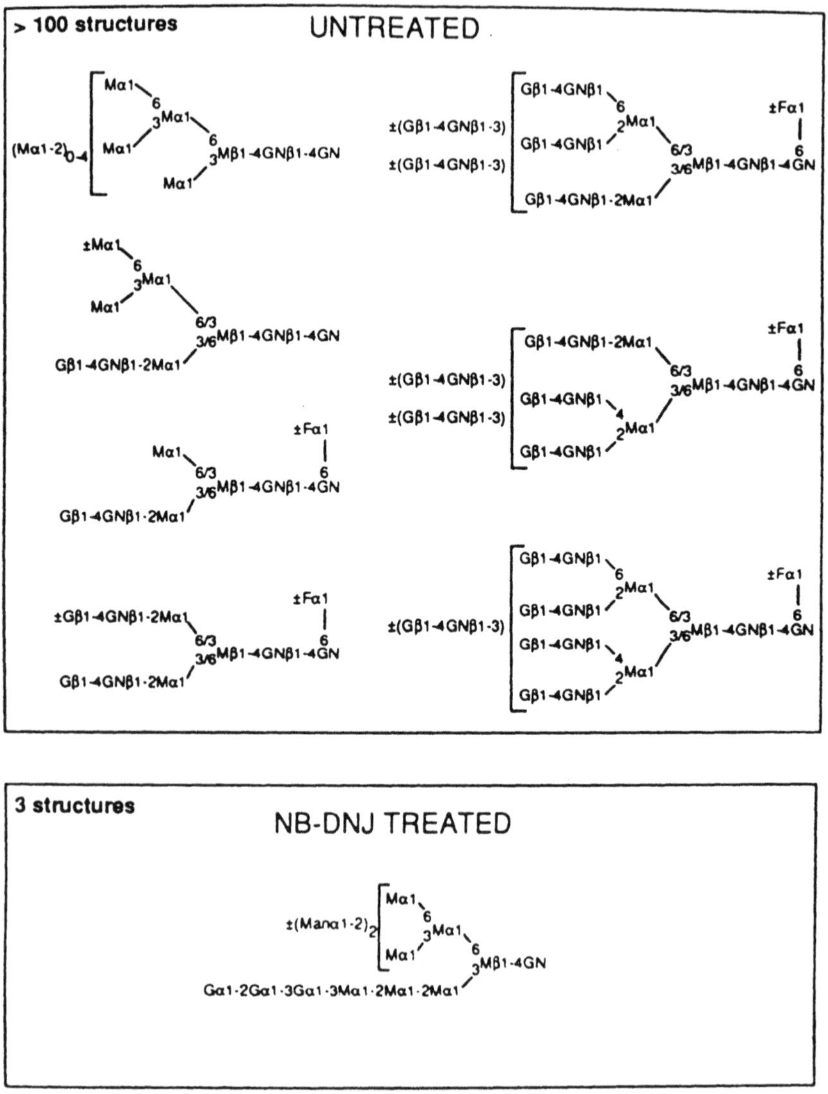

Figure 25. Summary of N-glycosylation structures found on untreated and NB-DNJ treated recombinant gp120 expressed in CHO cells. The upper panel shows the structures from untreated gp120 (Mizuochi et al., 1988b; Willey et al., 1988) and the lower panel shows the structures present following NB-DNJ treatment (Karlsson, 1993).

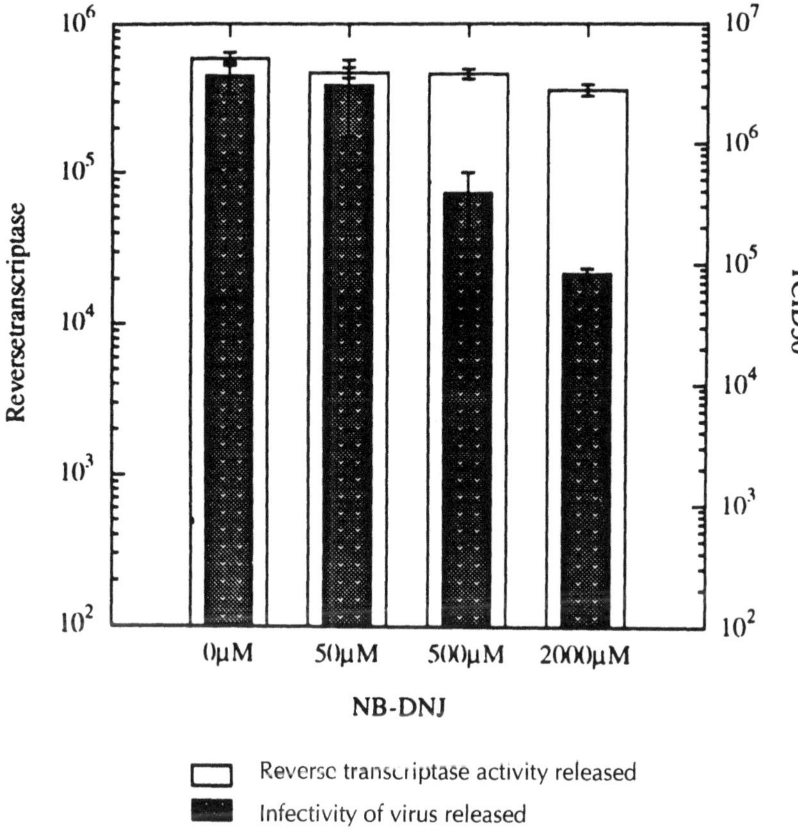

Figure 26. Effect of NB-DNJ on virus output and infectivity in H9 cells. Reverse transcriptase activity released (□). Infectivity of released virus (■) (P. Fischer, personal communication).

affords a possible therapeutic approach to managing glycolipid storage disorders by selectively altering cellular glycolipid levels to offset glucosylceramide accumulation (Platt et al., 1994).

SUMMARY

Clearly our current ability to determine glycoprotein and glycolipid structures, and to manipulate their biosynthetic pathways using specific glycosyltransferase and glycosidase inhibitors, has enormous therapeutic potential. In this way the importance of glycoproteins and glycolipids in normal development and morphogenesis can be further investigated and any disturbances resulting from disease potentially controlled.

ACKNOWLEDGMENTS

It is a privilege and honor to acknowledge my debt to the late Rodney Porter whose support led me into the field of immunology and then glycobiology. He encouraged me to form a relationship with the Monsanto Company to develop and automate the technology of microsequencing oligosaccharides. This led with the University of Oxford to the foundation of Oxford GlycoSystems, the first University spin-off company which was to develop technology arising mainly from within the Glycobiology Institute. I have benefited enormously from the help, advice and skills of my colleagues and students in developing the Glycobiology Institute and equipping it so as to be a resource for glycobiology. I thank the Biochemical Society for permission to reproduce figures from the 7th Wellcome Trust Award for Research in Biochemistry related to Medicine, 1994, published in the Biochemical Society Transactions, Volume 23.

REFERENCES

Alizon, M., Wain-Hobson, S., Montagnier, L., & Sonigo, P. (1986). Genetic variability of the AIDS virus: Nucleotide sequence analysis of two isolates from African patients. Cell 46, 63–74.

Amzel, L.M. & Poljak, R.A. (1979). Three-dimensional structure of immunoglobulins. Ann. Rev. Biochem. 48, 961–997.

Ashford, D.A., Alafi, C.D., Gamble, V.M., MacKay, D.J.G., Rademacher, T.W., Williams, P.J., Dwek, R.A., Barclay, A.N., Davis, S.J., Somoza, C., Ward, H.A., & Williams, A.F. (1993). Site-specific glycosylation of recombinant rat and human soluble CD4 variants expressed in Chinese hamster ovary cells. J. Biol. Chem. 268, 3260–3267.

Bode, W., Meyer, E., & Powers, J.C. (1989). Human leukocyte and porcine pancreatic elastase: X-ray crystal structures, mechanism, substrate specificity and mechanism-based inhibitors. Biochemistry 28, 1951–1963.

Brisson, J.-R. & Carver, J.P. (1983a). Solution conformation of asparagine-linked oligosaccharides: α(1-6)-linked moiety. Biochemistry 22, 3680–3686.

Brisson, J.-R. & Carver, J.P. (1983b). Solution conformation of asparagine-linked oligosaccharides: α(1-2)-, α(1-3)-, β(1-2)-, and β(1-4)-linked units. Biochemistry 22, 3671–3680.

Brockway, W. & Castellino, F.J. (1972). Measurement of the binding of antifibrinolytic amino acids to various plasminogens. Arch. Biochem. Biophys. 151, 194–199.

Burton, D.R., Boyd, J., Brampton, A.D., Easterbrook-Smith, S.B., Emanuel, E.J., Novotny, J., Rademacher, T.W., van Schravendijk, M.R., Sternburg, M.J.E., & Dwek, R.A. (1980). The C1q receptor site on Immunoglobulin G. Nature, Lond. 28, 338–344.

Bush, C.A., Blumberg, K., & Brown, J.N. (1982). Crystal structure and solution conformation of 1-N-acetyl-β-D-glucopyranosyl amine: A model for the glycopeptide linkage. Biopolymers 21, 1971–1977.

Carr, S.A., Hemling, M.E., Folena-Wasserman, G., Sweet, R.W., Anumula, K., Barr, J.R., Huddleston, M.J., & Taylor, P. (1989). Protein and carbohydrate structural analysis of a recombinant soluble CD4 receptor by mass spectrometry. J. Biol. Chem. 264, 21286–21295.

Datema, R., Olofosson, S., & Romero, P.A. (1987). Inhibitors of protein glycosylation and glycoprotein processing in viral systems. Pharmacology Therapeutics 33, 221–286.

Davis, S.J., Puklavec, M.J., Ashford, D.A., Harlos, K., Jones, E.Y., Stuart, D.I., & Williams, A.F. (1993). Expression of soluble recombinant glycoproteins with predefined glycosylation: Application to the crystallisation of the T-cell glycoprotein CD2. Protein Engineering 6, 229–232.

Davis, S.J., Ward, H.A., Puklavec, M.J., Willis, A.C., Williams, A.F., & Barclay, A.N. (1990). High level expression in Chinese hamster ovary cells of soluble forms of CD4 T-lymphocyte glycoprotein including glycosylation variants. J. Biol. Chem. 265, 10410–10418.

Deisenhofer, J., Enghild, J.J., Pizzo, S.V., & Gonzalez-Gronow, M. (1981). Crystallographic refinement and atomic models of a human Fc fragment and its complex fragment Ḃ of protein A from stapphylococcus aureus at 2.9- and 2.8-A resolution. Biochemistry 20, 2361–2370.

Delbaere, L.T.J. (1974). The molecular and crystal structures of 4-N-(2-acetamido-2-deoxy-β-D-glucopyranosyl)-L-asparagine trihydrate and 4-N-(β-D-glucopyranosyl)-L-asparagine monohydrate. Biochem. J. 143, 197–205.

Drickamer, K. (1988). Two distinct classes of carbohydrate-recognition domains in animal lectins. J. Biol. Chem. 263, 9557–9560.

Drickamer, K. & Taylor, M.E. (1993). Biology of Animal Lectins. Ann. Rev. Cell Biol. 9, 237–264.

Duncan, A.R. & Winter, G. (1988). The binding site for C1q on IgG. Nature, Lond. 332, 738–740.

Dwek, R.A., Ashford, D.A., Edge, C.J., Parekh, R.B., Rademacher, T.W., Wing, D.R., Barclay, A.N., Davis, S.J., & Williams, A.F. (1993a). Glycosylation of CD4 and Thy 1. Phil. Trans. Royal Soc. Lond. Biochem. 342, 43–50.

Dwek, R.A., Edge, C.J., Harvey, D.J., Wormald, M.R., & Parekh, R.B. (1993b). Analysis of glycoprotein-associated oligosaccharides. Ann. Rev. Biochem. 62, 65–100.

Ferguson, M.A.J. (1991). Lipid anchors on membrane proteins. Current Opinion in Structural Biology 1, 522–529.

Ferguson, M.A.J., Homans, S.W., Dwek, R.A., & Rademacher, T.W. (1988). Glycosyl-phosphatidylinositol moiety that anchors *Trypanosoma brucei* variant surface glycoprotein to the membrane. Science 239, 753–759.

Ferguson, M.A.J. & Williams, A.F. (1988). Inhibition of HIV replication by amino–sugar derivatives. Ann. Rev. Biochem. 57, 285–320.

Fleet, G.W.J., Karpas, A., Dwek, R.A., Fellows, L.E., Tyms, A.S., Petrusson, S., Namgoong, S.K., Ramsden, N.G., Smith, P.W., Son, J.C., Wilson, F., Witty, D.R., Jacob, G.S., & Rademacher, T.W. (1988). Inhibition of HIV replication by amino-sugar derivatives. FEBS Lett. 237, 128–132.

Freymann, D., Down, J., Carrington, M., Roditit, I., Turner, M., & Wilery, D. (1990). 2.9Å Resolution structure of the N-terminal domain of a variant surface glycoprotein from *Trypanosoma brucei*. J. Mol. Biol. 216, 141–160.

Fukuda, M. (1994). Cell surface carbohydrates: Cell-type specific expression. Molecular Glycobiology IRL Press, 1–52.

Gavel, Y. & Von Heijne, G. (1990). Sequence differences between glycosylated and non-glycosylated Asn-x-Thr/Ser acceptor sites: Implications for protein engineering. Protein Engineering 3, 433–442.

Geyer, H., Holschbach, C., Hunsmann, G., & Schneider, J. (1988). Carbohydrates of human immunodeficiency virus. J. Biol. Chem. 263, 11760–11767.

Gonzales-Gronow, M., Grennet, H.E., Fuller, G.M., & Pizzo, S.V. (1990). The role of carbohydrate in the function of human plasminogen: Comparison of the protein obtained from molecular cloning and expression in *Escherichia coli* and *cos* cells. Biochim. Biophys. Acta 1039, 269–276.

Gonzalez-Gronow, M., Edelberg, J.M., & Pizzo, S.V. (1989). Further characterization of the cellular plasminogen binding site: Evidence that plasminogen Z and lipoprotein A compete for the same site. Biochemistry 28, 2374–2377.

Goochee, C.F. & Monica, T. (1990). Environmental effects on protein glycosylation. Bio/Technology 8, 421–427.

Grinna, L.S. & Robbins, P.W. (1979). Glycoprotein biosynthesis. J. Biol. Chem. 254, 8814–8818.

Haltiwanger, R.S., Kelly, W.G., Roquemore, E.P., Blomberg, M.A., Dong, L.D., Kreppel, L., Chou, T., & Hart, G.W. (1992). Glycosylation of nuclear and cytoplasmic proteins is ubiquitous and dynamic. Biochem. Soc. Trans. 20, 264–269.

Harris, R.J., Chamow, S.M., Gregory, T.J., & Spellman, M.W. (1990). Synthesis and processing of asparagine-linked oligosaccharides. Eur. J. Biochem. 188, 291–300.

Harris, R.J., Leonard, C.K., Guzetta, A.W., & Spellman, M.W. (1991). Tissue plasminogen activator has an O-linked fucose attached to threonine-61 in the epidermal growth factor domain. Biochemistry 30, 2311–2314.

Harris, R.J. & Spellman, M.W. (1993). O-linked fucose and other post-translational modifications unique to EGF modules. Glycobiology 3, 219–224.

Hart, G.W. (1992). Glycosylation. Current Opinion in Cell Biology 4, 1017–1023.

Haurum, J.S., Arsequell, G., Lellouch, A.C., Wong, S.Y.C., Dwek, R.A., McMichael, A.J., & Elliot, T. (1994). Recognition of carbohydrate by major histocompatability complex class 1-restricted, glycopeptide-specific cytotoxic T lymphocytes. J. Exp. Med. 180, 739–744.

Hayes, M.L. & Castellino, F.J. (1979). Carbohydrate of the human plasminogen variants 1. Carbohydrate composition, glycopeptide isolation, and characterization. J. Biol. Chem. 254, 8768–8780.

Homans, S.W., Dwek, R.A., Boyd, J., Mahmoudian, M., Richards, W.G., & Rademacher, T.W. (1986). Conformational transitions in N-linked oligosaccharides. Biochemistry 25, 6342–6350.

Homans, S.W., Edge, C.J., Ferguson, M.A.J., Dwek, R.A., & Rademacher, T.W. (1989). Solution structure of the glycosylphosphatidylinositol membrane anchor glycan of *Trypanosoma brucei* variant surface glycoprotein. Biochemistry 28, 2881–2887.

Homans, S.W., Ferguson, M.A.J., Dwek, R.A., Rademacher, T.W., Anand, R., & Williams, A.F. (1988). Complete structure of the glycosylphosphatidylinositol membrane anchor of rat brain Thy-1 glycoproteins. Nature 333, 269–272.

Homans, S.W., Pastore, A., Dwek, R.A., & Rademacher, T.W. (1987). Structure and dynamics in oligomannose-type oligosaccharides. Biochemistry 26, 6649–6655.

Hubbard, S.C. & Ivatt, R.J. (1981). Synthesis and processing of asparagine-linked oligosaccharides. Ann. Rev. Biochem. 50, 555–583.

Ikeda, K., Sannoh, T., Kawasaki, N., Kawasaki, T., & Yamashina, I. (1987). Serum lectin with known structure activates complement through the classical pathway. J. Biol. Chem. 262, 451–454.

Joao, H.C. & Dwek, R.A. (1993). Effects of glycosylation on protein structure and dynamics in ribonuclease B and some of its individual glycoforms. Europ. J. Biochem. 218, 239–244.

Joao, H.C., Scragg, I.G., & Dwek, R.A. (1992). Effects of glycosylation on protein conformation and amide proton exchange rates in RNase B. FEBS Lett. 307, 343–346.

Jones, E.Y., Davis, S.J., Williams, A.F., Harlos, K., & Stuart, D.I. (1992). Crystal structure at 28A resolution of a soluble form of the cell adhesion molecule 2. Nature 360, 232–239.

Karlsson, G.B. (1993). D.Phil. Thesis. University of Oxford, Oxford, UK.

Karlsson, G.B., Butters, T.D., Dwek, R.A., & Platt, F.M. (1993). Effects of the imino sugar N-butyldeoxynojirimycin on the N-glycosylation of recombinant gp120. J. Biol. Chem. 268, 570–576.

Karpas, A., Fleet, G.W.J., Dwek, R.A., Petrusson, S., Namgoong, S.K., Ramsden, N.G., Jacob, G.S., & Rademacher, T.W. (1988). Amino sugar derivatives as potential anti-human immunodeficiency virus agents. Proc. Natl. Acad. Sci. USA 85, 9229–9233.

Kornfeld, R. & Kornfeld, S. (1985). Assembly of asparagine-linked oligosaccharides. Ann. Rev. Biochem. 54, 631–634.

Lasky, L.A., Groopman, J.E., Fennie, C.W., Benz, P.M., Capon, D.J., Dowbenko, D.J., Nakamuira, G.R., Nunes, W.M., Renz, M.E.B., & Berman, P.W. (1986). Neutralization of the AIDS retrovirus by antibodies to a recombinant envelope glycoprotein. Science 233, 209–212.

Leatherbarrow, R.J. & Dwek, R.A. (1983). The effect of aglycosylation on the binding of mouse IgG to staphylococcal protein A. FEBS Lett. 164, 227–230.

Leatherbarrow, R.J., Rademacher, T.W., Dwek, R.A., Woof, J.M., Clar, A., Burton, D.R., Richardson, N., & Feinstein, A. (1985). The effect of aglycosylation on the binding of mouse IgG to staphylococcal protein A. Mol. Immunol. 22, 407–415.

Lee, W.R., Yu, X.-F., Syu, W.-J., Essex, M., & Lee, T.-H. (1992). Mutational analysis of conserved N-linked glycosylation sites of human immunodeficiency virus type 1 gp41. J. Virol. 66, 1799–1803.

Leonard, C.K., Spellman, M.W., Riddle, L., Harris, R.J., Thomas, J.N., & Gregory, T.J. (1990). Assignment of intrachain disulfide bonds and characterisation of potential glycosylation sites of the type recombinant human immunodeficiency virus envelope glycoprotein (gp120) expressed in Chinese hamster ovary cells. J. Biol. Chem. 265, 10373–10382.

Lubas, W.A. & Spiro, R.G. (1987). Golgi endo-α-D-mannosidase from rat liver, a novel N-linked carbohydrate unit processing enzyme. J. Biol. Chem. 262, 3775–3781.

Lund, J., Tanaka, T., Takahashi, N., Sarmay, G., Arata, Y., & Jefferis, R. (1990). A protein structural change in aglycosylated human Ig correlates with loss of human Fcg RI and Fcg RIII binding and/or activation. Mol. Immunol. 27, 1145.

Malhotra, R., Lu, J., Holmskov, U.S., & Sim, R.B. (1994). Collectins, collectin receptors and the lectin pathway of complement. Clin. Exp. Immunol. 97 supp. 2, 4–9.

Malhotra, R., Wormald, M.R., Rudd, P.M., Fischer, P.B., Dwek, R.A., & Sim, R.B. (1995). Glycosylation changes of IgG associated with rheumatoid arthritis can activate complement via the mannose-binding protein. Nature Medicine 1, 237–243.

McPherson, A., Brayer, G., Cascio, D., & Williams, R. (1986). The mechanism of binding of a polynucleotide chain to pancreatic ribonuclease. Science 232, 765–768.

Mizuochi, T., Matthews, T.J., Kato, M., Hamako, J., Titani, K., Solomon, J., & Feizi, T. (1990). Diversity of oligosaccharide structures on the envelope glycoprotein gp120 of human immunodeficiency virus 1 from the lymphoblastoid cell line H9. Jn. Biol. Chem. 265, 8519–8524.

Mizuochi, T., Spellman, M.W., Larkin, M., Solomon, J., Basa, L.J., & Feizi, T. (1988a). Carbohydrate structures of the human-immunodeficiency-virus (HIV) recombinant envelope glycoprotein gp120 produced in Chinese hamster ovary cells. Biochem. J. (Japan) 254, 599–603.

Mizuochi, T., Spellman, M.W., Larkin, M., Solomon, J., Basa, L.J., & Feizi, T. (1988b). Structural characterization by chromatographic profiling of the oligosaccharides of human immunodeficiency virus (HIV) recombinant envelope glycoprotein gp120 produced in Chinese hamster ovary cells. Biomed. Chromatography 2, 260 270.

Mononen, I. & Karjalainen, E. (1987). Structural comparison on protein sequences around potential N-glycosylation sites. Biochem. Biophys. Acta 788, 364–367.

Morgan, B.P. (1990). Complement: Clinical Aspects and Relevance to Disease. Academic Press, London.

Mullin, N.P., Hall, K.T., & Taylor, M.E. (1994). Characterization of ligand binding to a carbohydrate-recognition domain of the macrophage mannose receptor. J. Biol. Chem. 269, 28405–28413.

Nishimura, H., Takao, T., Hase, S., Shimonishi, Y., & Iwanaga, S. (1992). Human factor IX has a tetrasaccharide O-glycosidically linked to serine 61 through the fucose residue. J. Biol. Chem. 276, 17520–17525.

Ny, T., Elgh, F., & Lund, B. (1984). The structure of the human tissue-type plasminogen activator gene: Correlation of intron and exon structures to functional and structural domains. Proc. Natl. Acad. Sci. USA 81, 5355–5359.

Parekh, R.B. (1987). A study of the structure and biosynthesis of N-linked oligosaccharides and of their involvement in certain diseases. D. Phil.Thesis, Oxford.

Parekh, R.B., Dwek, R.A., Sutton, B.J., Fernandes, D.L., Leung, A., Stanworth, D., Rademacher, T.W., Mizuochi, T., Taniguchi, K., Matsuta, K., Takeuchi, Y., Nagano, T., Miyamoto, T., & Kobata, A. (1985). Association of rheumatoid arthritis and primary osteoarthritis with changes in the glycosylation pattern of total serum IgG. Nature, Lond. 316, 452–457.

Parekh, R.B., Dwek, R.A., Thomas, J.R., Rademacher, T.W., Opdenakker, G., Wittwer, A.J., Howard, S.C., Nelson, R., Siegel, N.R., Jennings, M.G., Harakas, N.K., & Feder, J. (1989a). Cell-type-specific and site-specific N-glycosylation of type I and type II human tissue plasminogen activator. Biochemistry 28, 7644–7662.

Parekh, R.B., Isenberg, D.A., Ansell, B.M., Roitt, I.M., Dwek, R.A., & Rademacher, T.W. (1988). Galactosylation of IgG associated oligosaccharides: Reduction in patients with adult and juvenile onset rheumatoid arthritis and relation to disease activity. Lancet 1(8592), (April 30) 966–969.

Parekh, R.B., Isenberg, D.A., Rook, G., Roitt, I.M., Dwek, R.A., & Rademacher, T.W. (1989b). A comparative analysis of disease-associated changes in the galactosylation of serum IgG. J. Autoimmunity 2, 101–114.

Parekh, R.B., Tse, A.G.D., Dwek, R.A., Williams, A.F., & Rademacher, T.W. (1987). Tissue-specific N-glycosylation, site-specific oligosaccharide patterns and lentil lectin recognition of rat Thy-1. Eur. Mol. Biol. Org. J. 6, 1233–1244.

Patthy, L. (1985). Evolution of the proteases of blood coagulation and fibrinolysis by assembly from modules. Cell 41, 657–663.

Pirie-Shepherd, S., Jett, E.A., Andon, N.L., & Pizzo, S.V. (1995). Sialic acid content of plasminogen 2 glycoforms as a regulator of fibrinolytic activity. J. Biol. Chem. 270, 5877–5881.

Platt, F.M., Neises, G.R., Dwek, R.A., & Butters, T.D. (1994). N-butyldeoxynojirimycin is a novel inhibitor of glycolipid biosynthesis. J. Biol. Chem. 269, 8362–8365.

Ponting, C.P., Marshall, J.M., & Cederholm-Williams, S.A. (1992). Plasminogen: A structural review. Blood Coagulation Fibrinolysis 3, 605–614.

Rademacher, T.W., Edge, C.J., & Dwek, R.A. (1991). Dropping anchor with the lipophosphoglycans. Curr. Biol. 1, 41–42.

Rademacher, T.W., Homans, S.W., Parekh, R.B., & Dwek, R.A. (1986). Immunoglobulin G as a glycoprotein. Biochem. Soc. Symp. 51, 131–148.

Rademacher, T.W., Parekh, R.B., & Dwek, R.A. (1988a). Glycobiology. Ann. Rev. Biochem. 57, 785–838.

Rademacher, T.W., Parekh, R.B., Dwek, R.A., Isenberg, D., Rook, G., Axford, J., & Roitt, I. (1988b). The role of IgG glycoforms in the pathogenesis of rheumatoid arthritis. Springer Seminars in Immunopathology 10, 231–249.

Rao, U. & Teeter, M.M. (1993). Improvement of turn structure prediction by molecular dynamics: A case study of a1-purothionin. Protein Engineering 6, 837–847.

Rico, M. et al. (1989). Sequential 1H-NMR assignment and solution structure of bovine pancreatic ribonuclease. Ann. Euro. J. Biochem. 183, 623–638.

Rico, M. et al. (1991). 3D Structure of bovine pancreatic ribonuclease A in aqueous solution: An approach to tertiary structure determination from a small basis of 1H NMR NOE correlations. J. Biomolecular NMR 1, 283–298.

Robertson, A.D., Purisima, E.O., Eastman, M.A., & Scheraga, H.A. (1989). Proton NMR assignments and regular backbone structure of bovine pancreatic ribonuclease A in aqueous solution. Biochemistry 28, 5930–5938.

Roitt, I.M. et al. (1988). The role of antigen in autoimmune responses with special reference to changes in carbohydrate structure of IgG in rheumatoid arthritis. J. Autoimmunity 1, 499–506.

Rook, G.A.W. et al. (1991). Changes in IgG glycoform levels are associated with remission of arthritis during pregnancy. J. Autoimmunity 4, 779–794.

Rudd, P.M., Fortune, F., Patel, R.B., Dwek, R.A., & Lehner, T. (1994a). A human T cell receptor recognises "O" - linked sugars from the hinge region of human IgA1 and IgD. Immunology 83, 99–106.

Rudd, P.M. (1994b). Glycoforms modify the dynamic stability and functional activity of an enzyme. Biochemistry 33, 17–22.

Rudd, P.M. et al. (1994c). Glycoforms modify the dynamic stability and functional activity of an enzyme. Biochemistry 33, 17–22.

Santoro, J. et al. (1993). High-resolution three-dimensional structure of ribonuclease A in solution by nuclear magnetic resonance spectroscopy. J. Mol. Biol. 229, 722–734.

Schachter, H. (1986). Biosynthetic controls that determine the branching and microheterogeneity of protein-bound oligosaccharides. Biochem. Cell Biol. 64, 163–181.

Schachter, H. (1994). Molecular cloning of glycosyltransferase genes. In: Molecular Glycobiology, pp. 88–162. IRL Press.

Schachter, H. & Brockhausen, I. (1992). The biosynthesis of serine(threonine)-N-acetylgalactosamine-linked carbohydrate moieties. In: Glycoconjugates. Composition, Structure and Function, pp. 263–332. Marcel Dekker, New York.

Shaanan, B., Lis, H., & Sharon, N. (1991). Structure of a legume lectin with an ordered N-linked carbohydrate in complex with lactose. Science 254, 862–866.

Shall, S. & Barnard, E.A. (1969). Heavy atom-labelled derivatives of bovine pancreatic ribonuclease. I. Specific reactions of ribonuclease with N-acetylhomocysteine thiolactone and silver ion. J. Mol. Biol. 41, 237–251.

Sherrif, S., Chang, C.Y., & Ezekowitz, R.A.B. (1994). Human mannose-binding protein carbohydrate recognition domain trimerizes through a triple a-helical coiled-coil. Nature Struc. Biol. 11, 789–794.

Sottrup-Jensen, L., Claeys, H., Zajdel, M., Petersen, T.E., & Magnussen, S. (1978). Fibrinolysis and Thrombolysis, pp. 191–209. Raven Press, New York.

Spellman, M.W. et al. (1989). Carbohydrate structures of human tissue plasminogen activator expressed in Chinese hamster ovary cells. J. Biochemistry 264, 14100–14111.

Spellman, M.W., Leonard, C.K., Baa, L.J., Gelineo, I., & Van Halbeek, H. (1991). Carbohydrate structures of recombinant soluble human CD4 expressed in Chinese hamster ovary cells. Biochemistry 30, 2395–2406.

Sutton, B.J. & Phillips, D.C. (1983). The three dimensional structure of the carbohydrate within the Fc fragment of immunoglobulin G. Biochemical Soc. Trans. 11, 130–132.

Taylor, M.E. (1993). Recognition of complex carbohydrates by the macrophage mannose receptor. Biochemistry Soc. Trans. 21, 467–474.

Tsuchiya, N. et al. (1993). Detection of glycosylation abnormality in rheumatoid IgG using N-acetylglucosamine-specific Psathyrella velutina lectin. J. Immunology 151, 1137–1146.

Varki, A. (1993). Biological roles of oligosaccharides: All of the theories are correct. Glycobiology 3, 97–130.

Weis, W., Kahn, R., Fourme, R., Drickamer, K., & Hendrickson, W.A. (1991). Structure of the calcium-dependent lectin domain from a rat mannose-binding protein determined by MAD phasing. Science 254, 1608–1615.

Weis, W.I., Drickamer, K., & Hendrickson, W.A. (1992a). Structure of a C-type mannose-binding protein complexed with an oligosaccharide. Nature, Lond. 360, 127–134.

Weis, W.I. & Drickamer, K. (1994). Trimeric structure of a C-type mannose-binding protein. Curr. Biol. 2, 1227–1240.

Weis, W.I., Quesenberry, M.S., Taylor, M.E., Bezouska, K., Hendrickson, W.A., & Drickamer, K. (1992b). Molecular mechanisms of complex carbohydrate recognition at the cell surface. Cold Spring Harbor Symp. Quant. Biol. 57, 281–289.

Willey, R.L., Rutledge, R.A., Dias, S., Folks, T., Theodore, T., Buckler, C.E., & Martin, M.A. (1986). Identification of observed and divergent domains within the envelope of the acquired immunodeficiency syndrome retrovirus. Proc. Natl. Acad. Sci. USA 83, 5038–5042.

Willey, R.L. et al. (1988). *In vitro* mutagenesis identifies a region within the envelope gene of the human immunodeficiency virus that is critical for infectivity. J. Virology 62, 139–147.

Williams, A.F. & Gagnon, J. (1982). Neuronal cell Thy-1 glycoprotein: Homology with immunoglobulin. Science 216, 696–703.

Williams, A.F. et al. (1993). Comparative analysis of the N-glycans of rat, mouse & human Thy-1. Site-specific oligosaccharide patterns of neural Thy-1, a member of the immunoglobulin superfamily. Glycobiology 3, 339–348.

Williams, R.L., Greene, S.M., & McPhearson, A. (1987). The crystal structure of ribonuclease B at 2.5-Å resolution. J. Biol. Chem. 262, 16020–16031.

Winchester, B. & Fleet, G.W.J. (1992). Amino-sugar glycosidase inhibitors: Versatile tools for glycobiologists. Glycobiology 2, 199–210.

Wittwer, A.J. & Howard, S.C. (1990). Glycosylation at Asn-184 inhibits the conversion of single-chain to two-chain tissue-type plasminogen activator by plasmin. Biochemistry 29, 4175–4180.

Wittwer, A.J. et al. (1989). Effects of N-glycosylation on *in vitro* activity of bowes melanoma and human colon fibroblast derived tissue plasminogen activator. Biochemistry 28, 7662–7669.

Woods, R.J., Edge, C.J., & Dwek, R.A. (1994a). Protein surface oligosaccharides and protein function. Nature Structural Biology 1, 499–501.

Woods, R.J., Edge, C.J., Wormald, M.R., & Dwek, R.A. (1994b). GLYCAM 93: A generalized parameter set for molecular dynamics simulations of glycoproteins and oligosaccharides. Application to the structure and dynamics of a disaccharide related to oligomannose. In: Complex Carbohydrates in Drug Research. Alfred Benzon Symposium 36 (Bock, K. & Clausen, H., Eds.), pp. 15–26. Munksgaard, Copenhagen.

Woods, R.J., Fraser-Reid, B., Dwek, R.A., & Edge, C.J. (1993). The role of non-bonded interactions in determining the solution conformations of oligosaccharides. Modelling the Hydrogen Bond 569, 252–268.

Wooten, E.W., Bazzo, R., Edge, C.J., Zamze, S., Dwek, R.A., & Rademacher, T.W. (1990a). Primary sequence dependence of conformation in oligomannose oligosaccharides. Eur. Biophysics J. 18, 139–148.

Wooten, E.W., Edge, C.J., Bazzo, R., Dwek, R.A., & Rademacher, T.W. (1990b). Uncertainties in structural determination of oligosaccharide conformation using measurements of nuclear Overhauser effects. Carbohydrate Research 203, 13–17.

Wormald, M.R., Wooten, E.W., Bazzo, R., Edge, C.J., Feinstein, A., Rademacher, T.W., & Dwek, R.A. (1991). The conformational effects of N-glycosylation on the tailpiece from serum IgM. Euro. J. Biochem. 198, 131–139.

Yamashita, K. & Kamerling, J.P.A. (1982). Structural study of the carbohydrate moiety of hen ovomucoid. J. Biol. Chem. 257, 12809–12814.

Chapter 8

CELL CYCLES

J. Murdoch Mitchison

INTRODUCTION

Anyone writing a chapter about 30 years of modern science has a big field to cover. Murray and Hunt (1993, p. 14) in a recent attractive but narrowly focused book on the molecular biology of the cell cycle say, ". . . the key questions about the cell cycle were well defined by 1970. Yet the number of scientists investigating the cell cycle remained conspicuously small until the late 1980s." This is an illusion. Looking at books and long reviews during these 30 years tells a different story. At the start of this period Mazia[21] (1961) had more than 700 references, mainly but not exclusively about mitosis. My own files, a limited selection, showed the cell cycle papers per year rising steadily through the late 1950s and early 1960s to reach more than 100 per year in 1965 (Mitchison,[21] 1969). My book published 10 years after Mazia's (Mitchison, 1971) had more than 900 references. Prescott (1976) had 563 which were mostly different from mine. The long review by Hochhauser et al. (1981) had more than 1200 and the book by Lloyd et al. (1982) more than 2200. These are not the marks of a thinly populated field.

With limited space, I have often cited reviews rather than original papers, and, like any "outline" historian, I have omitted some topics such as the control of proliferation in mammalian cells and the nature of the G0 state (reviewed in Hochhauser et al., 1981; also in Baserga, 1985) and also the cell cycle of higher plants (reviewed in Yeoman, 1976). I have also not done full justice to other topics such as the bacterial cell cycle, mitosis, and circadian controls. In particular, I have only given a brief summary of one part of the modern molecular biology. This dominates present day research on the cell cycle, but it is well covered by modern reviews and I have assumed that most readers of a historical account will know the present state of the subject and will be more interested in older and less familiar stories. I have tried to give the flavor of the period and the debates that were happening, but my coverage is certainly incomplete. It is also personal, especially in the examples taken from fission yeast which I have worked with for nearly 40 years. Another historian would certainly tell a different story with different emphasis. I have also tried to point out some of the unsolved questions from the period in the hope that future work will be able to answer them.

In my view, progress in this field of biology was not primarily due to striking new ideas but rather to the facts that emerged, sometimes quite slowly, from new techniques such as the use of tritiated thymidine, the development of synchronous cultures, cell fusion, two-dimensional gels, and genetics, both classical and molecular.

G1, S, AND G2

In the early 1950s, the timing of DNA synthesis had been followed by microspectrophotometry of single cells stained by the Feulgen reaction. The most seminal method, however, was the use of autoradiography pioneered by Howard and Pelc (1953). They also named the phases of DNA synthesis during the cycle, now in universal use. S was the period of DNA synthesis preceded by the first gap (G1) and followed by the second gap (G2) before mitosis.

When tritiated thymidine became available in the late 1950s, DNA autoradiography really took off and it was one of the dominant techniques in cell cycle work throughout the 1960s, generating many papers on a wide variety of single cells, both free living and in tissues. Details of the methods were described in Mitchison (1971) but one widely used was to pulse label with tritiated thymidine and then follow the appearance of labeled mitoses in successive samples. This gave the lengths of G1, S, and G2 and had the advantage that it could be used in tissues where there were cells which were not cycling. It was also called a "retroactive" method (Prescott, 1976) since it did not depend on cell synchrony and the only handling of the culture was the pulse label given before the sampling.

There were several reasons for the popularity of DNA autoradiography:

1. DNA was recognized as a very important macromolecule.

2. The techniques were easy and cheap and did not require specialized equipment, although they could be time-consuming.
3. Thymidine was an efficient and specific label for newly synthesized DNA. Thymidine kinase, although not on the normal synthetic pathway, was present in most cells (but not all—e.g. yeasts).
4. Two important bits of cell biology made the analysis much easier. DNA doubled exactly from a IC value to a 2C one and, unlike most cell components, these values were invariant from cell to cell. Also the S period nearly always occupied a restricted part of the cycle. If it had been synthesized throughout the cycle, the method of scoring labeled mitoses could not have been used.

Most of the information about GI, S, and G2 in the 1960s came from DNA autoradiography. Flow cytometry with fluorescent DNA dyes is a later technique (reviewed by Melamed et al., 1979) which is now widely used. It is an excellent way of analyzing populations blocked in Gl or G2, and other parameters (e.g. RNA content or cell size) can be measured simultaneously with DNA content. It can also be used in growing populations provided all the cells are cycling. Its disadvantages are that the equipment is expensive and that it cannot be used on metazoan tissues.

DNA synthesis was also followed, usually with labeled thymidine, in synchronous cultures as they became available. An important paper by Helmstetter and Cooper (1968) using membrane elution (see below) showed that in fast growing cultures of *Escherichia coli* there was a constant time (C) for the replication of the circular chromosome. The surprising thing was that C could be longer than the whole cell cycle. The resolution of this paradox was that a round of replication could start before the previous one had finished.

By the end of the 1960s, it was clear that an important generalization could be made—that DNA synthesis was periodic in all cells except fast-growing prokaryotes. But it was not clear why this should be so and it remains a major unanswered question about the cell cycle. G2 can occupy three-quarters of the cell cycle in growing cells such as *Physarum* or fission yeast. It can be regarded as the time for "the preparations for mitosis," but this is a vague phrase like "preprophase" though it does recognize that the start of mitosis is difficult to define. We now know a lot more about the enzymes involved before mitosis, but we are still fairly ignorant about their kinetics of synthesis and activation. Do these kinetics occupy the long G2s of growing cells or is there some trigger in G2 like a "size control"? More is known now about the (old) "dependencies" or (new) "check point controls" which ensure that mitosis does not normally start until DNA replication has been completed, but this does not explain G2 since the minimum requirement for the cell is that mitosis should not precede the end of S.

Gl is equally or perhaps more puzzling. It has been known for some time that it is the most variable period of the cycle in mammalian and many other cells (Mitchison, 1971; Prescott, 1976). Does it involve a check point and, if so, what is being checked? One possibility is size or nucleo–cytoplasmic ratio. This is the case in budding yeast but it is much less clear in other cells. Are there specific Gl events or is it simply a period of growth (Cooper, 1991)? There is no clear evidence of GI-specific proteins except for those expressed in late Gl and needed for the Gl/S transition, but there are changes in chromosome condensation throughout Gl (see below). However, the most puzzling situations have always been those growing cells where there is no Gl. *Physarum* is a striking example where the S phase starts five minutes after telophase in a cycle which lasts 12 hours or more. *Amoeba proteus* is another example and perhaps some mammalian cells (Prescott, 1976; but see Brooks et al., 1983).

Tritiated thymidine was also used in the 1960s and 1970s to follow the events in the S period in finer detail. Unlike the bacterial chromosome, the eukaryotic chromosome initiated DNA synthesis in many places along its length though not all at the same time. In particular, there was evidence that heterochromatin was late-replicating. DNA synthesis was also followed at the level of individual DNA molecules by the "fiber autoradiography" technique set out in a classical paper by Cairns (1963). The detailed study by Huberman and Riggs (1968) on mammalian cells showed more than 10,000 "replicons" in which DNA synthesis proceeded bidirectionally from many origins.

It had been known for some time that the length of the S period varied considerably between the short cycles of most early embryos (except of mammals) and the longer cycles of growing somatic cells. In amphibian embryos, for example, the total DNA was synthesized 100 times faster than in adult somatic tissues. Surprisingly for many people, it turned out that the speed of movement of the replicating forks was much the same and the difference was that the short S periods were due to many more, but much shorter, replicons. Why there should be so many more origins of replication and how they are controlled is still mysterious.

In contrast, the circular chromosome of *E. coli* had only one origin. Much effort was spent in trying to find the direction of replication. By 1970, some five different methods (described in Mitchison, 1971) had produced what appeared to be "gas-tight" evidence for unidirectional replication. Yet Masters and Broda (1971) published, with some trepidation, evidence that replication was bidirectional, as in eukaryotes. Further experiments (see Cooper, 1991) showed that this was the right conclusion and that the earlier papers could be interpreted in other ways. This is the most striking example in the cell cycle field of the conclusion from a large body of evidence being overturned.

SYNCHRONOUS CULTURES

The earlier methods of cell cycle analysis described above used single cells and were very limited in the cell components that could be measured. For a proper biochemical analysis, it was necessary to use synchronous cultures in which cells divided at the same time and their cell cycles were aligned. Much effort was spent in the 1960s in developing and refining techniques for generating such cultures in a wide range of cells and they have been described in Mitchison (1971), Prescott (1976), and Lloyd et al. (1982). This was a key development in the history of the cell cycle and many hundreds of papers depended on the use of synchronous cultures. Without them, the analysis of the cell cycle would have been severely limited.

Induction Synchrony

In induction synchrony, a normal asynchronous cell culture is induced to divide synchronously by either physical or chemical changes in the environment. The earliest examples go back to the 1950s when Scherbaum and Zeuthen (1954) found that the ciliate *Tetrahymena* could be synchronized by short repetitive heat shocks, and Tamiya et al. (1953) induced synchronous division in the algal cells of *Chlorella* by light/dark cycles. At that time the generation of these synchronous cultures was in itself a fascinating phenomenon, even if the reasons for it were at first obscure and still have not been clarified at a molecular level. Later, Zeuthen and his colleagues developed the concept of "division proteins" which was one of the earliest control models (see below). The most popular technique, developed at first mainly for mammalian cells in the early 1960s, was the use of inhibitors to block cells for a time at a particular stage of the cycle, e.g. the S period or mitosis. The rationale was that all the cells would in due course reach this block point, and when the inhibitor was removed, they would then proceed synchronously through the cycle.

This block and release method still continues in use partly because it is easy and does not require special equipment and partly because it gives a very much higher yield of synchronized cells than the selection methods described below. New inhibitors were used—for instance α-factor, a pheromone which blocks budding yeast near the Gl/S boundary. A more recent development was the use of temperature-sensitive *cdc* mutants which block at various stages of the cycle after a shift to the restrictive temperature.

Apart from the general problems of synchronous cultures which are discussed below, there was a particular problem about induction synchrony which was realized early on but not always allowed for. During the block, cell growth may be affected but it does continue. On release, the oversize cells go through cell cycles which are shorter than normal and only later revert both to normal size and to normal times. There is therefore a distortion of the normal relations between growth and division, since division is faster than growth. This was shown in the original heat shock experiments on *Tetrahymena* by Scherbaum and Zeuthen (1954). A more recent analysis on fission yeast is in Novak and Mitchison (1990).

Another way of inducing synchronous cultures was to start with nongrowing stationary phase cells and then induce growth with fresh medium. This was partially successful with some prokaryotes and lower eukaryotes in the 1960s. Among the best examples is one for budding yeast which involved a series of alternating media and cell separations (Williamson and Scopes, 1961) and showed little distortion of the resulting synchronous culture. The equivalent technique for mammalian cells was much more widely used. Starting with quiescent cells, they were usually induced into growth by the addition of fresh medium with or without serum or growth factors. This was a fairly easy technique and, what is more, mimicked important situations in the whole animal like wound healing or tumor development. A major problem was to know how far the first cell cycle up to division (all that was often followed) is comparable to cell cycles in exponential growth. It involved passing through "restriction points" in Gl at which various components of the medium were needed, a reflection presumably of the complex growth controls in metazoan tissues. Most people would now agree that quiescent cells probably enter the normal cell cycle at the first S period, but the exact point is uncertain. There is recent evidence that gene expression is different in Gl entered from quiescence from what it is in cycling cells (Wick et al., 1994). Problems also arose because (not surprisingly) different mammalian cells have different cycles and apparently different controls. Many people in the past tended to work with one cell line with which they were familiar—human HeLa cells being probably the most popular over the last 40 years. But it became clear that transformed cells had different restriction points from primary cultures and the degree of difference varied with the extent of transformation. There were also differences between species since, for example, rodent lines responded to DNA inhibitors and other agents in a different way from human lines (Schimke et al., 1991). These differences have been a real complication with mammalian cell cycles in the past and one of the tasks for the future will be to explain them.

Selection Synchrony

Whereas induction methods synchronized all the cells of a culture, selection methods worked in a quite different way by physically selecting cells at a particular stage of the cycle in a normal asynchronous culture and then growing them up separately as a synchronous culture. This produced less distortion than the induction methods, but gave a much lower yield of synchronized cells.

The first of these selection techniques was developed by Terasima and Tolmach (1961), and has been called "selective detachment," "mitotic selection," or "wash off." Mammalian cells in monolayers round up at mitosis and lose their attachment to the substrate. Gentle shaking therefore produced a suspension of mitotic cells which were grown up separately. This was a powerful method which has been widely used since. The initial population comes from a narrow window of the cycle and is better synchronized than with most other methods.

A second method of selection was to separate a concentrated cell population by size (and density) after centrifugation in a sucrose gradient (Mitchison and Vincent, 1965). This copied the technique often used at the time for separating macromolecules. After velocity sedimentation for a few minutes, the top layer of small cells was separated off to make a synchronous culture. The method was shown to work with fission yeast, budding yeast, and *E. coli*. The degree of synchrony was not as good as with selective detachment because of the variation in cell size at the start of the cycle, but it could be, and was, used with many cell types.

Gradient separation has been modified in various ways since its introduction. Different non-metabolizable reagents were used to make the gradients. The yield was increased by scaling up the procedure in zonal rotors with gradients of 1.5 l. An important modification was to separate warm growing yeast cells by a counter current which balanced centrifugal force in an elutriating rotor (Creanor and Mitchison, 1979). This did not need gradients and eliminated certain perturbations that they caused.

An ingenious method of selecting young cells of *E. coli* was membrane elution (Helmstetter and Cummings, 1963). Medium was allowed to flow through a monolayer of cells which were packed tight on a membrane filter. As the cells divided, young daughter cells were released into the medium, presumably because there was not room for them on the membrane. These cells could be used to start synchronous cultures. But, since *E. coli* is sensitive to environmental changes and may be perturbed by the elution procedure (or by gradient separation), another "retroactive" method was developed by Helmstetter (1967). This involved treating the culture with a pulse of labeled thymidine *before* it was put on the membrane and eluted. The method has been described in detail by Cooper (1991), and led to the paper by Helmstetter and Cooper (1968), mentioned above. It was an excellent technique but it only worked with some bacterial strains.

We should pause briefly here to consider some of the limitations and criticisms that have been made about synchronous cultures since their introduction (e.g. Hochhauser et al., 1981; Cooper, 1991). Induction methods involve the distortion of the relations between growth and division mentioned above. In addition, some of the inhibitors used have side effects apart from their blocking action. *cdc* mutants are "safer," but they involve temperature shifts. Repetitive temperature changes could give misleading results (e.g. Lark and Maaloe, 1956, showing discontinuous DNA synthesis in a prokaryote). Selection methods are better, but there are also possibilities of distortion. For example, gradient selection could produce perturbations probably caused by concentrating the cells initially or by the effects of the gradient material and the change of medium. However, there are fundamental problems about all synchronous cultures. The most important is that their synchrony is imperfect since the cells do not all divide at the same time. In selection synchrony, some of this is due to imperfect selection, but even if the selection is good, as in selective detachment, the normal variation in cell cycle traverse time will produce imperfect division synchrony (Prescott, 1976) and this will get worse in later cycles. The effect of this on a "peak pattern" (e.g. a cyclin) is to spread out the peak and reduce its height. Methods have been developed to correct for this but they have seldom been used (Creanor and Mitchison, 1982, 1994). Even then, all that emerges is the pattern for the *average* single cell. It conceals the variation between individual cells which is only revealed by single-cell observations. For example, single growing cells of fission yeast have a change in the rate of length growth which on average is about mid-G2. But there is a large variation in its position in G2 which would make it undetectable in a synchronous culture (Mitchison and Nurse, 1985). These problems in synchronous cultures have made it difficult to analyze some of the fine details of the cycle, and there seems at present no easy way of solving them.

Age Fractionation

Cells in a gradient are fractionated by size and, in principle, by cell age in the cycle. Successive samples down the gradient should therefore resemble successive samples in a synchronous culture. This technique started in the mid-1960s and is described in some detail in Lloyd et al. (1982). It had two great advantages. It used all the cells in a culture and so had a high yield. It could also be used retroactively since the whole culture could be labeled with a tracer and then cooled rapidly (and treated with an inhibitor such as cycloheximide) before being age fractionated on the gradient. It was widely used on budding yeast and is still in use for mammalian cells. Elutriation rotor separation of chilled cells was later used as an alternative to gradients.

The problem of this method was that the fractionation of the larger cells was not very efficient, perhaps because density altered as well as size. There was a second problem that came in relating the samples to the stage in the cycle. In an ideal world, cell size should double over the cycle. But in fact the cell size varied by nearly threefold because of the size variation at any one stage of the cycle. What then should be done with the tails of the size distribution? This difficult problem, which can profoundly alter the apparent cell cycle pattern, is discussed in Creanor et al. (1983), and J.M. Mitchison (1988).

Natural Synchrony

Early embryos have nearly perfect natural synchrony for a number of cycles after fertilization. They have a long and distinguished history over more than a century for the cytological study of mitosis and division (Mazia, 1961). The most widely used were echinoderm eggs (sea urchins and starfish) which could be obtained in quantities sufficient for biochemical analysis (an early example is Rapkine, 1931). Amphibian eggs, especially those of *Xenopus laevis*, were also an important material since they were large enough for easy microinjection. In the mid-1980s, cell-free extracts of amphibian eggs were developed and proved a powerful way of overcoming the barrier of the cell membrane to external agents. Another important egg was that of *Drosophila* (see below under Genetics).

These eggs are highly specialized cells with very short cycles (down to 10 min) in which there is little growth apart from DNA and a few proteins, and little if any Gl and G2. This had the advantage of separating off and simplifying the events of mitosis, cleavage, and the S period, but it missed out the controls exercised by growth in all normal cells. An exception, both with growth and with precise natural synchrony, is the plasmodium of the myxomycete (slime mold) *Physarum polycephalum*. It is a syncytium without cell membranes between its many nuclei and it became a popular material with the development of axenic cultures around 1960 with its enthusiastic backing by H.P. Rusch. It played an important part in cell fusion experiments and in other cell cycle work. I have 136 reprints on *Physarum*, by no means a complete collection, yet, somewhat surprisingly, it is not mentioned in the recent book by Murray and Hunt (1993).

GROWTH AND ENZYME SYNTHESIS

In principle, all growing cells double their components during the cell cycle. An early interest was to define the patterns of growth. Was it continuous, or periodic like DNA synthesis? Was it exponential or were there rate changes? This work started in the mid-1950s and was mostly done on single cells which avoided the problems of poor synchrony in the later synchronous cultures. Since the best technique was to use single living cells, the measures of growth were restricted in most cases to optical ones such as volume, or dry mass by interferometry. An early and striking exception was the use by Prescott (1955) of a Cartesian diver balance to measure the "reduced weight" (approximately equivalent to total dry mass) in a single growing *Amoeba*. This was a technical *tour de force* and also showed a growth curve in which the rate of increase of reduced weight steadily decreased through the cycle to reach nearly zero before division. The exquisite sensitivity of the Cartesian diver respirometer was exploited much later by Hamburger et al. (1977) to measure the CO_2 production in single fission yeast cells.

These single-cell studies tapered off in the mid-1960s as synchronous cultures were used more and more. Reviews can be found in Mitchison (1971) and Prescott (1976), and they emphasize the variety of patterns that were found in different cells. In many but not all cases, growth in volume and dry mass were continuous. In some cases, it was exponential with an increasing rate, but in others it was "linear" with a constant rate through most of the cycle and a sharp doubling in rate at one point. In a few cases, there was a decreasing rate through the cycle as in *Amoeba* or *Streptococcus*. It is worth remembering that in all cases except exponential growth there was a sharp change in growth rate which was cell cycle related and implied major changes in metabolism. However, these patterns have not been explained and are now largely forgotten.

In the later 1960s, there was a great burst of activity in measuring enzyme synthesis, or more strictly enzyme activity, in synchronous cultures [Mitchison (1971) listed 77 papers from this period]. It was an attractive and relatively easy task to apply standard biochemical enzyme assays to samples from synchronous cultures. Undoubtedly the question in people's minds was whether continuous patterns of growth or total protein synthesis concealed periodic patterns of synthesis of individual proteins at different points in the cycle. If it happened with DNA in eukaryotes, why should it not happen with proteins?

What emerged was a variety of patterns, mostly in bacteria and lower eukaryotes. There were "step enzymes" in which the activity doubled fairly sharply (like DNA) at some point in the cycle. There were "peak enzymes" in which the activity rose sharply and then declined as the enzyme was presumably degraded or inactivated. Some enzymes also showed a pattern of continuous rise in activity. In some cases, this continuous rise showed a linear pattern, though it needed careful and frequent measurements to be sure of this.

The first model for the control of enzyme synthesis can be called "oscillatory repression" and was developed primarily for bacteria. There was an excellent review on this subject by Donachie and Masters (1969). Fully repressed enzymes were synthesized continuously (sometimes following a linear pattern) but partially repressed enzymes showed a series of steps once per cycle. Oscillations could have been produced by end-product repression and such stable oscillations were called "autogenous". An explanation of their cell cycle timing was that they were entrained by gene replication even though oscillations could continue in the absence of DNA synthesis. This was a persuasive model at the time and there was support for it in the eukaryote *Chlorella* (Molloy and Schmidt, 1970). However, it was difficult to test critically and little further work was done on it in the 1970s and beyond.

The second model was put forward in the mid-1960s for budding yeast by Halvorson and his colleagues. It was called "sequential transcription" or "linear reading" and was extensively reviewed by Halvorson et al. (1970). Some 30 enzymes all showed step changes in activity at various points in the cell cycle and the model was that these steps were caused by sequential transcription of the structural genes along all or part of the genome. There was a considerable amount of supporting evidence in an organism with a well-mapped genome. For instance, the order of the steps followed the order of four enzyme genes located on the fifth chromosome. Again, Cox and Gilbert (1970) used two strains of budding yeast in which two enzyme genes on the second chromosome were separated by different amounts. The same difference was found in the activity steps.

Some work continued on budding yeast, including a novel way of measuring the enzyme activity of single cells (Yashphe and Halvorson, 1976) and the prevalence of step enzymes was emphasised in a short review by Halvorson (1977). But the position was different in fission yeast where there had also been extensive enzyme assays in synchronous cultures (Mitchison, 1977). Only one out of 19 enzymes examined showed a step pattern. The others showed continuous increases in activity, in some cases following a linear pattern.

This apparent conflict was puzzling, and there was increasing worry that the process of synchronizing budding yeast could cause perturbations which might produce step patterns, and also that analysis of age fractionation was not altogether straightforward. Using the less perturbing elutriation technique, Creanor et al.

(1983) did not find the steps in three budding yeast enzymes which had appeared in earlier synchronous cultures.

Whatever was thought about the apparent conflict in yeasts, the most influential results came from a new technique. This was the development of high resolution two-dimensional gel electrophoresis by O'Farrell (1975), which could resolve several hundred individual proteins in samples from synchronous or age-fractionated cultures. Elliott and McLaughlin (1978) analyzed 111 of the more abundant proteins in budding yeast and found that they were all synthesized continuously. In a more extensive study of budding yeast by Lorincz et al. (1982), only 17 out of about 900 proteins were not synthesized continuously. Similar results were found in *E. coli* by Lutkenhous et al. (1979), and in HeLa cells by Bravo and Celis (1980).

This seemed the end of the story as far as the *abundant* proteins (including the major "housekeeping" metabolic enzymes) were concerned. As Nasmyth (1994), said: "Growth, which is boringly continuous and therefore harder to study, has been widely ignored." True enough, though linear patterns in fission yeast were followed in a small way through the 1980s (Mitchison, 1989). But some additional comments should be made. Step enzymes were found by activity measurements and activity does not necessarily reflect synthesis. Histones were also found to be synthesized periodically during the S period and enzymes concerned with DNA synthesis often showed peak patterns of activity. So also did the enzymes of mitosis (see below) where a key component, the p34 kinase, was activated by posttranslational modification. The difference between periodic enzymes in budding yeast and continuous ones in fission yeast surfaced again in White et al. (1986). Both the activity of DNA ligase and its mRNA were periodic in budding yeast and continuous in fission yeast. Finally sequential transcription emerged again in the homeotic genes in development. The *HOX* genes of vertebrates are turned on sequentially in time and in space and the temporal sequence follows the order of the genes on the chromosome (Duboule, 1994). This "colinearity" remains a puzzle.

Another way of using enzyme assays was to test for inducibility or "potential" by inducing an enzyme in samples taken from a synchronous culture and measuring the rate of increase of activity. In bacteria, there was a stepwise increase in this rate at a time when the enzyme gene was replicated (Donachie and Masters, 1969). This was regarded as an example of gene dosage and could in principle be used to map the genome. But, unlike bacteria, fission yeast showed steps in "potential" when DNA synthesis was blocked, so gene dosage did not apply here (Benitez et al., 1980).

This section has largely been concerned with enzymes but these were by no means the only cell components that were followed in synchronous cultures. There were many papers on cell cycle changes in other components such as nonenzyme proteins, lipids, small molecules, ions, and gas exchange, but there is not space here to outline the results. Some of them are discussed in Hochauser et al. (1981) and Lloyd et al. (1982).

CONTROL MODELS

Two important developments in the 1970s were in genetics (described below) and in experimentally based models for the control of mitosis and DNA synthesis. The flavor of this period of model building is well set out in the chapters in John (1981).

Division Proteins

The earliest model for division control was developed by Zeuthen, Scherbaum, and their colleagues in the early 1960s primarily to explain why repetitive heat shocks synchronize the division of the ciliate *Tetrahymena* (reviewed in Mitchison, 1971). One or more proteins were synthesized throughout the cycle and formed a stable structure at the "transition point" towards the end of the cycle. This structure was responsible for division. Heat shocks before the transition point caused a breakdown of the proteins or an intermediate unstable structure, and the cells were "set back" and had to start again, thus delaying division. Heat shocks after the transition point did not affect the stable structure or delay division. This was a neat way of explaining the synchronizing effect. Happily, there was a protein structure, the oral apparatus, which behaved in the same way as the postulated division proteins. The division proteins have never been identified though there has been recent suggestion that one of them might be a cyclin (Williams and Macey, 1991).

Size Control

It is an old idea that cells only divide when they have reached a critical size. For instance, Hartmann (1928) found that periodical amputation of the cytoplasm of *Amoeba* stopped division, presumably because the critical size was never reached. This type of size control was widely discussed in the 1970s with a distinction drawn between "sizers" where an event or period of time in the cycle was altered by size and "timers" where they were not.

As often happened, modern work started with *E. coli* where a short theoretical paper by Donachie (1968) showed that DNA replication started when the cell mass per unit chromosome origin reached a critical level. Within a range of cycle times, there was then a constant time until division. In effect, there was a sizer followed by a timer. Although a lot more is now known about the genes and proteins of *E. coli*, the link between initiation and size has not yet been discovered (Donachie, 1993).

In fission yeast, Fantes (1977), using time-lapse films of individual cells, showed that in normal populations cell size altered cycle time but not growth rate. Large cells had short cycle times and small cells had long cycle times. This provided a homeostatic mechanism for maintaining mean cell size and implied a size control. With oversize cells produced by a *cdc* block, the size control was thought not to operate and division happened after running through an "incompressible G2" (a timer). These results did not locate where cell size was monitored, but later experiments using nutritional shifts and the temperature sensitive mutant *weel-50* were interpreted as showing that the critical size was reached near mitosis. In the case of the small *wee* mutants, the mitotic size control was inactivated and replaced by another size control operating near the Gl/S boundary which was cryptic in wild-type cells. These two size controls, only one of which was operative in any one strain, was a sophisticated model at the time. Reviews of these topics are in Fantes and Nurse (1981) and Nurse and Fantes (1981). There was a known genetic element, *weel*$^+$, in the mitotic size control but otherwise the nature of both size controls were and still are mysterious.

At about the same time, experiments on budding yeast, somewhat less extensive than those on fission yeast, showed that a critical size had to be reached before "Start" (the commitment to the cell cycle in Gl). This situation (Carter, 1981) was therefore similar to *wee* mutants in fission yeast. This critical size, as in fission yeast, varied with the nutritional conditions. It was assumed at the time that this was the major control point. Subsequently it turned out, in a rather confused situation, that some mutants revealed a second control point near mitosis, as in fission yeast, which *might* be affected by cell size (e.g. Veinot-Drebot et al., 1991). It is worth making the point that when there is an early size control around a critical size in Gl, as in normal budding yeast, *size at division* is affected not only by the critical size but also by growth during the timer period. Nutritional changes can affect both the sizer and the timer.

It is not certain whether there is a size control in mammalian cells. Some experiments, going back to the mid 1960s, supported the idea of a size control near the Gl/S boundary but a few did not (reviewed briefly in Fantes and Nurse, 1981). It would be fair to say that the jury is still out. But interest in this situation has largely subsided and a definitive verdict may be long delayed.

The last of the important papers on sizers and timers analyzed the cell cycles of the algal cells *Chlamydomonas* and *Chlorella* in which a long period of growth is followed by rapid multiple S periods and divisions (Donnan et al., 1985; McAteer et al., 1985). A timer, temperature compensated in the range 20–30 °C, ran from the start of the cycle to a point of commitment in late Gl. It was followed by another timer, relatively insensitive to growth. However, during the period of this second timer, there were a number of rounds of S periods and divisions controlled by sizers which determined how many of these rounds occurred—the larger the cell, the more rounds. In a short review, John (1984) points out the similarities with other cell types.

It is sad that recent molecular biology has thrown so little light on the nature of these size controls. They are important for two reasons. First, they provide a mechanism for coordinating growth with the periodic events of DNA synthesis and mitosis, ensuring cell size homeostasis. Second, the Gl sizers (which are the majority) are early events in the cycle which act as triggers for the first periodic event of DNA synthesis. The G2 sizers may equally well be a trigger for the biochemical changes which precede mitosis. Murray and Hunt (1993) regard size controls as a brake on the "cell cycle engine," but they equally well be may be the starter. Perhaps we suffer a little from the recent concentration on early embryos where size controls clearly cannot operate in any simple sense.

Cell Fusion

Starting in the mid-1960s and continuing vigorously through the 1970s, a series of experiments exploited the natural synchrony of the plasmodia of *Physarum* and the ease of plasmodial fusion to explore the initiation of mitosis (reviewed in Sachsenmaier, 1981). A key experiment was to fuse an early G2 plasmodium with a late G2 one. The nuclei in the fused plasmodium all divided synchronously with the early G2 nuclei being accelerated and the late G2 nuclei being delayed in a dose-dependent way.

The model that came from these experiments was a "titration" one in which an initiator protein (or "mitogen") was synthesized continuously throughout G2 and adsorbed onto nuclear binding sites. When these sites were saturated, there was a sharp rise in the cytoplasmic concentration of the initiator which led to mitosis. At mitosis, the binding sites doubled and the whole process repeated in the next cycle. The initiator was assumed to be unstable in order to explain the delay in some nuclei, and parallels were drawn with the instability of the "division proteins" of Zeuthen and of the instability implied by the Hartmann experiment on *Amoeba*. Cell size was regulated by assuming the initiator was made at a rate proportional to cell mass.

A few years later, the development of virus-induced fusion led to parallel experiments with mammalian cells (reviewed in Rao and Sunkara, 1978). The timings could not be made with the same precision as in *Physarum* but the systems did allow the study of events in Gl, a phase lacking in *Physarum*. Fusion of cells in different stages of Gl indicated that an S-phase initiator was generated steadily through Gl, like the mitotic initiator in the G2 of *Physarum*. Fusion of Gl cells with G2 cells showed that the S-phase initiator did not act on G2 cells in which there was no further DNA synthesis. Fusion of Gl cells with mitotic cells caused premature chromosome condensation in the Gl cells. The degree of condensation decreased throughout GI suggesting that chromosomal changes were real events in Gl and that it was not simply a period of waiting.

In recent years, molecular biology has produced some candidates for these initiators. Moreno et al. (1990) suggest that *cdc25* is a mitotic initiator in fission yeast, whereas Nasmyth (1993) regards different cyclins as initiators of both mitosis and S-phase in budding yeast. Whether these correspond to the initiators postulated from the fusion experiments remains to be seen. It will be necessary to look carefully at the levels, timing, and location of these proteins.

Transition Probability

Most models of cell cycle control were and still are deterministic. A sequence of events is thought to occur which control the passage through the cycle. If variability, for example of timing, is considered at all, it is thought to be due to the sum of random processes round the events. In the early 1970s, however, an alternative model for cycle control was suggested, with the initial evidence primarily from films of mammalian cells (reviewed in Hochhauser et al., 1981). In its simplest form, cells in early G1 were in the A-state, a kind of limbo, out of which they exited at a transition point to the B-state with a defined probability. The B-state included a deterministic sequence finishing at mitosis. Most of the variation in cycle time came from the probability of the transition from the A- to the B-state.

This model became an important and controversial topic in the 1970s with detailed studies of the intermitotic times of sister cells and of unrelated cells. It also became more complicated with two transition points which gave a better fit to the data. This is reviewed in Brooks (1981) with an interesting discussion about cell size. But interest in the model flagged in the 1980s partly because no molecular or cytological event could be identified at the transition point(s) and partly because the model was found not to work well for slow-growing cells (Brooks and Riddle, 1988). Even so, it should be remembered as a serious attempt to explain variability, a fact of the cell cycle which is often swept under the carpet.

Sequences and Parallel Pathways

If the cell cycle followed a simple deterministic sequence then stopping one event should also stop subsequent events. Such a sequence is DNA synthesis followed by mitosis and division. These dependency relations, nowadays called "check point controls," are an active field of research at present. However, it is also possible that they may be two or more sequences or parallel pathways which converge towards the end of the cycle. If two events are on different pathways then blocking one of them may not block the other.

Parallel pathways to mitosis were mentioned explicitly in Mazia (1961). They were also suggested in Mitchison (1971) and developed further in Mitchison (1974). Two possible pathways were the "DNA-division cycle" (the sequence above) and the "growth cycle". The growth cycle was more shadowy but was put forward largely because inhibitor studies showed that growth could continue after a block to the DNA-division cycle. I suggested that step and peak enzymes could be markers in growth. This turned out to be wrong but nevertheless there proved to be markers in growth in fission yeast defined by the changes in rate in linear patterns (Mitchison, 1989).

Parallel pathways played an important role in the genetic analysis of the cycle since *cdc* mutants provided a far wider range of blocks than inhibitors. A complex plan of multiple pathways in budding yeast is in Pringle and Hartwell (1982).

GENETICS AND MOLECULAR BIOLOGY

The genetics of the cell cycle started first with *E. coli*. In the mid-1960s, workers at the Pasteur Institute isolated a large series of temperature-sensitive mutants which affected DNA synthesis and cell division (Hirota et al., 1968). It turned out that many of them affected metabolism and the SOS repair system for DNA damage. However, work continued in defining cell cycle genes and their products in much the same way as in eukaryotes. In recent reviews by Donachie (1992, 1993) perhaps the most striking conclusion is that only a few gene products are *specifically* required to carry out the periodic events of the bacterial cycle—a contrast to eukaryotes. But it is fair to say that the genetic approach in bacteria did not attract the same publicity as it did in eukaryotes, probably because of the basic differences in DNA replication, chromosome separation, and cell division.

Hartwell was the pioneer in eukaryotic cell cycle genetics. In 1969 he discovered and named *cdc* (cell division cycle) mutants in budding yeast, and began what he later described as, "The most exciting time in my scientific career" (Hartwell, 1993). These mutants were temperature-sensitive conditionals and, at the restrictive temperature, blocked progress through the cycle mostly at particular morphological stages. In an influential review, Hartwell et al. (1974) laid out the "circuitry" of the cell cycle with 19 genes needed to progress through 9 stages of the cycle. Most of these stages formed a dependent sequence, but bud emergence and nuclear migration were on a separate parallel pathway. These dependent sequences amplified and extended the "DNA-division cycle" where the dependencies mostly came from inhibitor blocks. *cdc* genes were more numerous and more specific. There was also evidence of a "timer" control for bud emergence. An important conclusion was that *CDC28* was needed early in the cycle and defined the commitment to progress through the cycle. This point in the cycle was called "Start," a name now in general use although Hartwell (1993) regrets it, no doubt because it can be confusing. Another name coined in this period was the "execution point." This was when a cell cycle gene had completed its function, so a mutant of this gene if shifted to the restrictive temperature after this point, would continue to the end of the cycle. Operationally, it is the same as the transition point (see above) and this latter name was used with *cdc* mutants in fission yeast.

Following the success of Hartwell's work, Nurse et al. (1976) isolated and characterized 14 *cdc* genes in fission yeast which blocked at four stages of the cycle. The initial screen was mostly for cells longer than normal at the restrictive temperature, exploiting the fact that growth continued after a *cdc* block, as it also did in budding yeast. $cdc2^{+}$ was unusual since in its absence the cell cycle was blocked at two stages, before mitosis and in Gl. During the 1980s, its gene product $p34^{cdc2}$ was identified as a protein kinase which played a key role before mitosis.

Genetic work on yeast in the later 1970s (reviewed in Simchen, 1978) continued with ordering gene effects and dependencies, for example by finding the terminal phenotypes of double mutants and by "reciprocal shifts." Analysis of revertants and suppressors was also informative. More mutants continue to be isolated and Murray and Hunt (1993, Appendix) listed for budding yeast the formidable totals of 54 *cdc* mutants and 229 other mutants affecting the cell cycle. For fission yeast, the comparable figures were 25 and 91.

Nor was genetic analyses of the cell cycle restricted to yeasts and bacteria. Simchem (1978) reviewed them in *Aspergillus*, *Chlamydomonas*, and *Tetrahymena*, and Murray and Hunt (1993) listed those for *Aspergillus*. Starting in the 1970s, cell cycle mutants were also isolated in various mammalian cells, a slightly puzzling result in diploid lines (Simchen, 1978; Lloyd et al., 1982; Baserga et al., 1985).

Another material which was exploited from the mid-1980s was the early embryo of *Drosophila* with its powerful genetic background. Glover (1989) listed some 70 genes that play a role in mitosis, some of which had homologues in yeast. This early embryo has an interesting development which is different from that of amphibian and echinoderm eggs. There is an initial syncytium with very rapid cycles and labile controls, e.g. the centrosome cycle can be dissociated from mitosis. Later the controls become tighter as cellularization develops. A recent review is by White-Cooper and Glover (1995).

The most important development in the 1980s was the application of the new techniques of molecular biology or molecular genetics. Genes could be identified from "libraries," cloned, and sequenced. As the data bases of gene sequences rapidly expanded, it became increasingly easy to find homologues that might define their function. Genes could be disrupted and site-specific mutations became possible. In some organisms, foreign genes could be introduced to test for functional complementation, and also integrated into the genome. With hybrid vectors, genes could be expressed in bacteria, so enabling an antibody to be raised without an initial chemical purification of the protein. Perhaps the major limitation was that the physiological substrate for an enzyme was difficult to identify.

These powerful new techniques came to dominate the field as they did with many other aspects of biochemistry and cell biology. Two symposia volumes, John (1981) and Brooks et al. (1989), illustrate the change during the decade. However, for reasons I gave in the Introduction, I will only give a brief summary of one particularly important discovery of these times. Much more detailed reviews (which also list other reviews) are in Nurse (1990), Forsburg and Nurse (1991), Marsh (1992), and Murray and Hunt (1993). The Cold Spring Harbor Symposium on Quantitative Biology Vol. 5 (1991) has 83 papers on the cell cycle, mostly about its molecular biology.

The most important discovery of the 1980s was the regulatory network that preceded mitosis (reviewed in Murray and Hunt, 1993). Physiological studies starting in the early 1970s showed that frog oocytes could be stimulated into meiosis by a "maturation promoting factor" (MPF). In the mid-1980s, MPF was also shown to induce mitosis in the cells of early embryos. Another important discovery of this period was the existence of cyclin, a protein made from stored maternal mRNAs in early embryos, which had the unusual property of being built up during interphase and then breaking down rapidly at mitosis. Its importance in the cell cycle was shown clearly in frog egg extracts where all the stored mRNAs had been destroyed yet adding purified cyclin mRNA alone would stimulate the early stages of mitosis.

Meanwhile, work on fission yeast had shown that a key player in mitosis was the protein kinase p34^{cdc2} the product of *cdc2*$^{+}$ (whose homologue in budding yeast was *CDC28*). It was activated posttranslationally by dephosphorylation of a tyrosine residue. Mitosis was advanced by the tyrosine phosphatase *cdc25* and delayed by the protein-kinase *wee1*.

The separate work on eggs and fission yeast came together in the late 1980s with the biochemical purification of MPF—a mammoth task. It seemed likely from molecular weights that the MPF was a complex of cyclin B and p34, but definitive proof came from recognition by antibodies. The discovery of these two key proteins, their association in a complex, and the activation by changes in phosphorylation were rightly hailed as a major success of molecular genetics and biochemistry in understanding the control of the cell cycle. These were new molecules whose existence was not even suspected in the early days. What is more, they were also identified in a wide range of eukaryotic cells, though not in *E. coli*. It seemed that the system was widely conserved and in this sense "universal." This is not surprising since mitosis itself is wellnigh universal in eukaryotes.

Typically perhaps, the situation began to get more complicated in the late 1980s. More cyclins were identified in budding yeast and were involved in the control of Start and premitotic phases of the cell cycle. The same was true of mammalian cells where there was a rapidly growing family of cyclins and also cyclin-dependent kinases (cdk) where the kinases differed either slightly or more conspicuously from p34^{cdc2}. This was, and still is, a fast moving but somewhat confused field.

Another problem has been that only the middle part of the control pathway was identified. Although the p34 kinase could be assayed by its phosphorylation of a foreign Hl histone, it was not clear what its substrate was *in vivo*. So the end of the pathway was unknown.

The beginning of the pathway was also a problem, not perhaps in the very rapid cycles of early embryos where cyclin accumulation started at the end of mitosis. Even so, Hinegardner et al. (1964) had found that prophase in sea urchin eggs started about a third of a cycle before the beginning of anaphase. This was before the peak of cyclin accumulation and of active MPF. But it was more of a problem in growing cells. When and why did "these preparations for mitosis" (to use an old phrase) start? In most cases "when" was difficult to judge since timing was relatively unimportant in the molecular models. "Why" was equally difficult. If it was a size control, the molecular basis of this was unknown. These problems were there in 1990 and they still remain a challenge for the future.

MITOSIS AND CYTOKINESIS

These are dramatic and visible splitting events which have interested cell biologists for more than a century, thus having a much longer history than any of the synthetic events described earlier in this chapter. But they mainly involve structure and biophysics rather than biochemistry and were always treated rather separately from the synthetic events. Although they are essential parts of the cell cycle, they were omitted from some of the major books and reviews on the cycle. Those who worked on them often went to symposia, and contributed to the resulting symposia volumes on subjects like motility, microtubules, or the cytoskeleton rather than the cell cycle.

Mitosis

Early work on mitosis was reviewed at length by Mazia (1961). Since then, there have been many reviews of which a small sample are those by Niklas (1971, 1988), Dustin (1978), McIntosh (1982), T.J. Mitchison (1988), McIntosh and Hering (1991), and Desai and Mitchison (1995).

The mitotic spindle was recognized early on as a key structural element in mitosis. Polarized light microscopy showed that it was birefringent and so possessed submicroscopic structure. In the 1950s, the electron microscope showed that this structure could be resolved into a fibrous array of "microtubules" running between the centrioles at the poles of the spindle and the kinetochores on each chromosome. By the early 1970s, the main chemical components were identified as two proteins, α and β tubulin, though there were also other proteins associated with them. An important part of the identification of the tubulin was the discovery that colchicine bound specifically to it.

The next major advance in the fine structure of the spindle came in the early 1980s, mainly from the work of J.R. McIntosh and his colleagues. This showed that the microtubules had polarity with plus and minus ends. The minus ends were those adjacent to the centriole and the plus ends were either at the kinetochores, or they interdigitated in the center of the spindle with other plus ends from the opposite pole. The plus ends were more active and had higher rate constants for assembly and dissociation *in vitro*.

Coupled with this was the increasing evidence in the 1970s and early 1980s that the spindle was not a fixed structure but a dynamic one with continuous turnover of the microtubule elements. Spindle microtubules had a turnover time of 15–50 seconds—much faster than that of interphase microtubules. But stabilization depended on position and time in mitosis.

At the start of our period, it was reasonable to regard the kinetochore as a simple anchor for the attachment of a spindle fiber. But by the late 1960s, it was becoming clear that this was too simple a view. The inventory of the kinetochore proteins increased through the 1980s and beyond and included "motor" proteins (dynein and kinesin-like), though their exact function is still not understood. However, "Kinetochores are clearly complex and wonderful machines" (McIntosh and Hering, 1991).

A final structural component of the spindle is the centriole, containing in many cells a centrosome. This is certainly an initiator of microtubules and its component molecules were becoming known in the late 1980s when centrosomes could be separated in biochemical quantities from mammalian cells and budding yeast. As with kinetochore proteins, the exact way they work is not yet known.

The structure and dynamics of the spindle are obviously important, but the fundamental fascination has been how chromosomes are moved and separated. There was certainly progress here during our period even though Taylor (1975) pointed out how slow this progress was compared to that with muscle and flagella. But chromosome movement is different since it is not a rapidly repetitive process and the forces and velocities involved are strikingly small (Niklas, 1988).

The first part of chromosome movement is "congression" when the chromosomes move to the metaphase plate and the chromatid pairs become attached to microtubules running to the spindle poles. An old problem was why each chromatid attached to opposite poles. By the end of the 1960s, elegant microdissection experiments on meiotic bivalents provided partial explanation. The only stable situation was when a pair was attached to opposite poles and under apparent tension. If a pair became attached to one pole only, this attachment was unstable and the mistake was rapidly corrected. But the details of this process were (and still are) poorly understood. For instance, it was not clear why the pairs appeared to be pushed away from the poles and became equidistant between them.

The second chromosome movement is in anaphase when the pairs separate and appear to be under a poleward force with the kinetochores leading. Various suggestions were made in the 1960s and 1970s about the mechanisms involved. One of them invoked "traction fibers" which were separate from a stable framework of microtubules and which were not birefringent. This fell out of favor in the 1980s because it became clear that the framework was not stable and because the fibers could not be identified. A second model (dating back to 1929) involved sliding filaments with a rough analogy to muscle or flagella. The earlier versions in the late 1960s were in time rejected, but more sophisticated models continued into the 1990s with emphasis on cross bridges or, later on, motor enzymes, causing microtubules to slide over each other.

Experiments in the early 1950s and extended in the mid-1980s led to models which emphasized the importance of disassembly of microtubules as a major mechanism in anaphase movement. In the simplest form of these models, shortening of microtubules by disassembly would drag a simple kinetochore anchor towards the pole. It was shown that disassembly can do work, but, as Dustin (1978) pointed out, it is difficult to see how disassembly alone could produce a directed force. It is like pulling yourself up by your own bootstraps. Something else was needed and it was very likely to be in or near the kinetochore where the major disassembly was known to occur (T.J. Mitchison, 1988). Exactly what this is will not be known until the structure and function of kinetochores becomes clearer.

By 1990, the consensus opinion was that microtubule assembly and disassembly played a key role in chromosome movements, but sliding of microtubules was not ruled out. Several recent reviews have suggested that more than one mechanism may be at work either normally or as backups. It is also worth remembering that although in broad outline mitosis is a universal process in eukaryotes, the details and perhaps the balance of mechanisms may differ, especially in lower eukaryotes.

Cytokinesis

Cytokinesis, the final process in the cell cycle, was somewhat of a Cinderella during our period. Part of the reason for this is that it lacked the attraction of a universal mechanism such as mitosis in eukaryotes. Animal cells divide with a medial cleavage furrow. Prokaryotes and fungi such as yeast form an ingrowing septum, whereas higher plants form a phragmoplast and then a cell plate growing outwards. In the latter cases, there was progress in describing the fine structure and in isolating genes which affected cytokinesis, but much less in understanding its mechanism. This is not surprising since morphogenesis and directed growth were, and still are, mysterious subjects. It is worth mentioning, however, the discovery of the preprophase band of cortical microtubules which appeared at the position of the future cell plate in higher plants but then disappeared at mitotic prophase (Gunning, 1982). Its function was unknown.

Most attention was focused on animal cells, especially in eggs where the cleavage furrow is large and conspicuous. Some of the more recent reviews for our period are in Rappaport (1986), Mabuchi (1986), and Satterwhite and Pollard (1992).

By 1960, there were a number of theories of the mechanism of cytokinesis which had been developed in the preceding 20 years (reviewed in Swann and Mitchison, 1958). One of them, which later became the dominant explanation, was a contracting ring in the cleavage furrow, but others invoked a stimulus at the polar ends of the cell which caused expansion or relaxation of the surface, or a direct involvement of the mitotic apparatus. It is worth remembering that at that time a contracting ring had not been visualized nor was it easy to imagine then how it could contract to nothing in the final stages of cleavage.

Progress in the 1970s and 1980s came from a series of observations and experiments. It became clear that the mechanism of cleavage resided in the cortex or cell surface rather than the mitotic apparatus since this apparatus could be removed or destroyed without stopping the cleavage process. Even fragments of the equatorial cortex could show furrowing.

Better electron microscopy around 1970 showed a ring of microfilaments in the furrow region which were later identified as actin. It also became clear that this ring was a dynamic structure (like the spindle) which could disassemble. The bulk isolation of cleavage furrows produces an inventory (still probably incomplete) which included myosin II as well as actin together with other proteins which could act on actin, e.g. by severing or cross-linking (Mabuchi, 1986).

The elegant and important microdissection experiments of Rappaport over more than 20 years showed that there was a "cleavage stimulus," a surface-acting factor in eggs which moved from the mitotic axis in anaphase to the site of the furrow at

6–8 μm/min. He also listed 15 experiments which supported the idea of a furrow stimulus rather than a polar stimulus (Rappaport, 1986).

By 1990, the consensus opinion was that cytokinesis in animal cells was carried out by the contraction of an equatorial ring in the furrow composed of actin and myosin. What was not known was how actin attached to the surface, became organized into a ring, and then disappeared as the furrow contracted. Nor was it known how myosin accumulated in the furrow nor what was the nature of the cleavage stimulus. Surface growth also remained a problem since an egg increases its surface by 26% at cell division. Was the surface pulled out passively and, if so, when did it recruit new material?

OSCILLATORS

The existence and function of oscillators in the cell cycle was a continuing topic during our period both for theoreticians and for experimentalists. An early example, mentioned above, was oscillatory feedback repression as the cause of stepwise increases in enzyme activity. Tyson (1979, 1983) emphasized the constraints on such a model and also discussed why the period of oscillation should be close to the cell cycle time.

A vigorous debate in the 1970s and 1980s centered around the control of mitosis in *Physarum*. The titration model with nuclear binding sites was in mathematical terms an extreme relaxation oscillator with saw-tooth kinetics. On the other side was the concept of a sinusoidal oscillator with a limit cycle and a point of singularity where the oscillations stopped. These two points of view were well developed by Tyson and Sachsenmaier (1984) and Shymko et al. (1984). Although limit cycle oscillators were the dominant concept for "clocks" such as the circadian timekeeper, the experimental evidence for the relaxation oscillator was better in the case of *Physarum*.

A cytoplasmic oscillator or "clock" emerged from the work on surface contractions in early embryos. In *Xenopus* embryos, there is a contraction which tends to round up the cell profile at the time of division. Hara et al. (1980) found that these contractions continued in enucleate fragments or in cells treated with anti-mitotic drugs. Similar results were also found with sea urchin eggs. So there was a cytoplasmic oscillator which continued to function in the absence of DNA synthesis or mitosis. Later work indicated that this oscillator was the periodic accumulation and breakdown of MPF (a relaxation oscillator) and that it was independent of its downstream effects on nuclear events (Murray and Hunt, 1993). It was not quite as simple as that since the careful studies of Shinagawa (1983) showed that the period of the *Xenopus* oscillator lengthened by about 30% after enucleation or treatment with colchicine, so there was some connection between the oscillator and mitosis.

Another case of oscillator control was found in growth events in fission yeast (reviewed in Mitchison, 1989 and in Mitchison et al., 1991). A number of events of growth, e.g. the rate of CO_2 production, were found to follow a linear pattern with a sharp doubling at one point in the cell cycle (not always the same point with different events). The identification of these periodic rate change points (RCP) made it possible to answer the question raised from the early model of the growth cycle. When growth continued after a block to the DNA-division cycle with a *cdc* mutant, did the periodic RCPs also continue? The answer was yes, though only for a limited time. So there was an oscillatory control which could function independently of nuclear events. What was more surprising was that the period of these persistent oscillators varied between different growth events, thus there was no "master" oscillator controlling growth after a block. Instead, these oscillators appeared to be entrained and synchronized in normal growth by some event of the DNA-division cycle. There was some success in identifying which event it might be, but the molecular basis of this synchronization was unknown and it remained a problem for future.

Biological rhythms or chronobiology is a large subject with a literature that is at least as great as that of the cell cycle. But there is an overlap with the cell cycle that I shall briefly consider here. A much fuller account is in the articles in the book edited by Edmunds (1984).

It became clear in the early 1950s that an important generator of rhythms was a circadian oscillator which was widely present in eukaryotes at many levels of organization from cells to whole animals. This sinusoidal oscillator had the following properties: (1) it could be synchronized or entrained to a precise 24 h period by light or temperature cycles; (2) it could be predictably phase-shifted by single light or temperature signals; (3) under constant environmental conditions it would "free run" with a persistent period near to 24 h; and (4) this free-running period was temperature compensated, i.e. invariant over a physiological range of temperature. These two latter properties are those of a real clock and are unlike the cell cycle. Good evolutionary reasons can be given for the development of such a clock.

The cell cycles in the tissues of higher animal and plant cells are controlled by a range of external hormones and other factors, and it was not surprising that a clock located somewhere in an animal should affect the cell cycles in its tissues. It was more surprising that a circadian oscillator should exist in unicellular organisms including nonphotosynthetic ones such as yeasts and ciliates (for a list see Edmunds and Laval-Martin, 1984).

The clearest cases of interaction between a circadian oscillator and the cell cycle came from situations where the cell cycle was longer than 24 h. As an example, in an early paper by Sweeney and Hastings (1958) on the marine dinoflagellate *Gonyaulax polyedra*, cultures were synchronized by light/dark cycles and then kept in constant dim light at 18.5 °C. During the constant light period, the average cycle time was 4.2 days but there was a burst of division every 23.9 h, interpreted as the period of the free-running circadian oscillator. Nine bursts of division were observed with little damping. Only some of the cells divided at each burst, probably the larger ones (Homma and Hastings, 1989), and individual cells would have cycle times which were integral multiples of 23.9 h. This phenomenon has been called "gating" of cell division to a time determined by the circadian oscillator, and has been observed in other organisms.

A second situation is where the cell cycle time had been adjusted to be near 24 h and the cycle was thought to be entrained by the circadian oscillator. An example is in Jarrett and Edmunds (1970) where a single dark-to-light transition induced partial synchronization of division in *Euglena* which lasted for five cycles in the light without damping. Control of division by a circadian oscillator may well occur here, but there is a general problem of which control is dominant when both the cell cycle and the circadian oscillator have much the same period. It has not always been possible to find both the two diagnostic features of a free-running circadian oscillator—persistence without damping and temperature compensation. It is even more difficult when these two features are only present partially.

The third situation is where the cell cycle time is much shorter than 24 h. Here there has been no evidence of input from a circadian oscillator. There was some evidence of relatively short period oscillations mainly in metabolic events (see Chapters 1, 4, and 6 in Lloyd et al., 1982), but their connection with the cell cycle remained obscure.

A suitable conclusion is the last sentence of Murray (1989): "It seems likely that we still have a great deal to learn about timekeeping and coordination in the cell cycle."

ACKNOWLEDGMENTS

I am grateful to W.D. Donachie, D.M. Glover, J.R. McIntosh and T.J. Mitchison for discussion and correspondence.

REFERENCES

Baserga, R. (1985). The Biology of Cell Reproduction. Harvard University Press, Cambridge, MA.

Benitez, T., Nurse, P., & Mitchison, J.M. (1980). Arginase and sucrase potential in the fission yeast *Schizosaccharomyces pombe*. J. Cell Sci. 46, 399–431.

Bravo, R. & Celis, J.E. (1980). A search for differential polypeptide synthesis throughout the cell cycle of HeLa cells. J. Cell Biol. 84, 795–802.

Brooks, R.F. (1981). Variability in the cell cycle and the control of cell proliferation. In: The Cell Cycle (John, P.C.L., Ed.), pp. 35–61. Cambridge University Press, Cambridge.

Brooks, R., Fantes, P., Hunt., T., & Wheatley, D., Eds. (1989). The cell cycle. J. Cell Sci. Suppl. 12.

Brooks, R.F. & Riddle, P.N. (1988). The 3T3 cell cycle at low proliferation rates. J. Cell Sci. 90, 601–612.

Brooks, R.F., Riddle, P.N., Richmond, N., & Marsden, J. (1983). The G1 distribution of "Gl-less" V79 Chinese hamster cells. Exp. Cell Res. 148, 127–142.

Cairns, J. (1963). The bacterial chromosome and its manner of replication as seen by autoradiography. J. Molec. Biol. 6, 208–213.

Carter, B.L.A. (1981). The control of cell division in *Saccharomyces cerevisiae*. In: The Cell Cycle (John, P.C.L., Ed.), pp. 99–117. Cambridge University Press, Cambridge.

Cooper, S. (1991). Bacterial Growth and Division. Academic Press, San Diego.

Cox, C.G. & Gilbert, J.B. (1970). Nonidentical times of gene expression in two strains of *Saccharomyces cerevisiae* with mapping differences. Biochem. Biophys. Res. Commun. 38, 750–757.

Creanor, J. & Mitchison, J.M. (1982). Patterns of protein synthesis during the cell cycle of the fission yeast *Schizosaccharomyces pombe*. J. Cell Sci. 58, 263–285.

Creanor, J. & Mitchison, J.M. (1994). The kinetics of Hl histone kinase activation during the cell cycle of wild-type and *wee* mutants of the fission yeast *Schizosaccharomyces pombe*. J. Cell Sci. 107, 1197–1204.

Creanor, J., Elliott, S.G., Bisset, Y.C., & Mitchison, J.M. (1983). Absence of step changes in activity of certain enzymes during the cell cycle of budding and fission yeast in synchronous cultures. J. Cell Sci. 61, 339–349.

Desai, A. & Mitchison, T.J. (1995). A new role for motor proteins as couplers to depolymerising microtubules. J. Cell Biol. 128, 1–4.

Donachie, W.D. (1968). Relationship between cell size and time of initiation of DNA replication. Nature, Lond. 219, 1077–1079.

Donachie, W.D. (1992). What is the minimum number of dedicated functions required for a basic cell cycle? Curr. Opin. Genet. Devel. 2, 792–798.

Donachie, W.D. (1993). The cell cycle of *Escherichia coli*. Annu. Rev. Microbiol. 47, 199–230.

Donachie, W.D. & Masters, M. (1969). Temporal control of gene expression in bacteria. In: The Cell Cycle. Gene-Enzyme Interactions (Padilla, G.M., Whitson, G.L., & Cameron, I.L., Eds.), pp. 37–76. Academic Press, New York.

Donnan, L., Carville, E.P., Gilliland, T.J., & John, P.C.L. (1985). The cell cycles of *Chlamydomonas* and *Chlorella*. New Phytol. 99, 1–40.

Duboule, D. (1994). Temporal colinearity and the phyllotypic progression: a basis for the stability of the vertebrate Bauplan and the evolution of morphologies through heterochrony. Development, Suppl. 135–142.

Dustin, P. (1978). Microtubules. Springer-Verlag, Berlin.

Edmunds, L.N., Ed. (1964). Cell Cycle Clocks. Marcel Dekker, New York.

Edmunds, L.N. & Laval-Martin, D.L. (1984). Cell division cycles and circadian oscillators. In: Cell Cycle Clocks (Edmunds, L.N., Ed.), pp. 295–324. Marcel Dekker, New York.

Elliott, S.G. & McLaughlin, C.S. (1978). Rate of macromolecular synthesis through the cell cycle of the yeast *Saccharomyces cerevisiae*. Proc. Natl. Acad. Sci. USA 75, 4384–4388.

Fantes, P.A. (1977). Control of cell size and cycle time in *Schizosaccharomyces pombe*. J. Cell Sci. 24, 51–67.

Fantes, P.A. & Nurse, P. (1981). Division timing: controls, models and mechanisms. In: The Cell Cycle (John, P.C.L., Ed.), pp. 11–33. Cambridge University Press, Cambridge.

Forsburg, S.L. & Nurse, P. (1991). Cell cycle regulation in the yeasts *Saccharomyces cerevisiae* and *Schizosaccharomyces pombe*. Annu. Rev. Cell Biol. 7, 227–256.

Glover, D.M. (1989). Mitosis in *Drosophila*. J. Cell Sci. 92, 37–146.

Gunning, B.E.S. (1982). The cytokinetic apparatus: its development and spatial regulation. In: The Cytoskeleton in Plant Growth and Development (Lloyd, C.W., Ed.), pp. 229–292. Academic Press, London.

Halvorson, H.O. (1977). A review of current models on temporal gene expression in *Saccharomyces cerevisiae*. In: Cell Differentiation in Microorganisms, Plants and Animals (Nover, L. & Mothes, K., Eds.), pp. 361–376. VEB Gustav Fischer Verlag, Jena.

Halvorson, H.O., Carter, B.L.A., & Tauro, P. (1970). Synthesis of enzymes during the cell cycle. Adv. Microbiol. Physiol. 6, 47–106.

Hara, K., Tydeman, P., & Kirschner, M. (1980). A cytoplasmic clock with the same period as the division cycle in *Xenopus* eggs. Proc. Natl. Acad. Sci. USA 77, 462–466.

Hartmann, M. (1928). Uber experimentelle unsterblichkeit von protozoen-individuen. Ersatz der fortpflanzung von *Amoeba proteus* durch fortgesetzte regeneration. Zool. Jahrbucher. 45, 973–987.

Hartwell, L.H. (1993). Getting started in the cell cycle. In: The Early Days of Yeast Genetics (Hall, M.N. & Linder, P., Eds.), pp. 307–314. Cold Spring Harbor Press, Cold Spring Harbor.

Hartwell, L.H., Culotti, J., Pringle, J.R., & Reid, B.J. (1974). Genetic control of the cell cycle in yeast. Science 183, 46–51.

Helmstetter, C.E. (1967). Rates of DNA synthesis during the division cycle of *Escherichia coli B/r*. J. Molec. Biol. 24, 417–427.

Helmsetter, C.E. & Cooper, S. (1968). DNA synthesis during the division cycle of rapidly growing *Escherichia coli B/r*. J. Molec. Biol. 31, 507–518.

Helmetetter, C.E. & Cummings, D.J. (1963). An improved method for the selection of bacterial cells at division. Biochim. Biophys. Acta 82, 608–610.

Hinegardner, R.T., Rao, B., & Feldman, D.E. (1964). The DNA synthetic period during the early development of the sea urchin egg. Exp. Cell Res. 36, 53–61.

Hirota, Y., Ryter, A., & Jacob, F. (1968). Thermosensitive mutants affected in the processes of DNA synthesis and cellular division in *Escherichia coli*. Cold Spring Harbor Symp. Quant. Biol. 33, 677–693.

Hochhauser, S.J., Stein, J.L., & Stein, G.S. (1981). Gene expression and cell cycle regulation. Internat. Rev. Cytol. 71, 95–243.

Homma, K. & Hastings, J.W. (1989). Cell growth kinetics, division asymmetry and volume control at division in the marine dinoflagellate *Gonyaulax polyedra*: a model of circadian clock control of the cell cycle. J. Cell Sci. 92, 303–318.

Howard, A. & Pelc, S.R. (1953). Synthesis of desoxyribonuclei acid in normal and irradiated cells and its relation to chromosome breakage. Heredity, London Suppl. 6, 261–273.

Huberman, J.A. & Riggs, A.D. (1968). On the mechanism of DNA replication in mammalian chromosomes. J. Molec. Biol. 32, 327–341.

Jarrett, R.M. & Edmunds, L.N. (1970). Persisting circadian rhythms of cell division in a photosynthetic mutant of *Euglena*. Science 167, 1730–1733.

John, P.C.L., Ed. (1981). The Cell Cycle. Cambridge University Press, Cambridge.

John, P.C.L. (1984). Control of the cell division cycle in *Chlamydomonas*. Microbiol. Sci. 1, 96–101.

Lark, K.G. & Maaloe, O. (1956). Nucleic acid synthesis and the division cycle of *Salmonella typhimurium*. Biochim. Biophys. Acta 21, 448–458.

Lloyd, D., Poole, R.K., & Edwards, S.W. (1982). The Cell Division Cycle. Academic Press, New York.

Lorincz, A.T., Miller, M.J., Xuong, N.H., & Geiduschek, E.P. (1982). Identification of proteins whose synthesis is modified during the cell cycle of *Saccharomyces cerevisiae*. Molec. Cell Biol. 2, 1532–1549.

Lutkenhaus, J.F., Moore, B.A., Masters, M., & Donachie, W.D. (1979). Individual proteins are synthesized continuously throughout the *Escherichia coli* cell cycle. J. Bacteriol. 138, 352–360.

Mabuchi, I. (1986). Biochemical aspects of cytokinesis. Internat. Rev. Cytol. 101, 175–213.

Marsh, J., Ed. (1992). Regulation of the Eukaryotic Cell Cycle. Ciba Foundation Symp. 170. John Wiley, Chicester.

Masters, M. & Broda, P. (1971). Evidence for the bidirectional replication of the *Escherichia coli* chromosome. Nature, New Biol. London. 232, 137–140.

Masui, Y. (1992). Towards understanding the control of the division cycle in animal cells. Biochem. Cell Biol. (Canada) 70, 920–945.

Mazia, D. (1961). Mitosis and the physiology of cell division. In: The Cell (Brachet, J. & Mirsky, A.E., Eds.), Vol. 3, pp. 77–412. Academic Press, New York.

McAteer, N., Donnan, L., & John, P.C.L. (1985). The timing of division in *Chlamydomonas*. New Phytol. 99, 41–56.

McIntosh, J.R. (1982). Mitosis and the cytoskeleton. In: Developmental Order: Its Origin and Regulation (Subtelny, R. & Green, P.B., Eds.), pp. 79–115. Alan R. Liss, New York.

McIntosh, J.R. & Hering, G.E. (1991). Spindle fiber action and chromosome movement. Annu. Rev. Cell Biol. 7, 403–426.

Melamed, M.R., Mullaney, P.F., & Mendelsohn, M.L. (1979). Flow Cytometry and Sorting. Wiley, New York.

Mitchison, J.M. (1969). Markers in the cell cycle. In: The Cell Cycle. Gene-Enzyme Interactions (Padilla, G.M., Whitson, G.L., & Cameron, I.L., Eds.), pp. 361–372. Academic Press, New York.
Mitchison, J.M. (1971). The Biology of the Cell Cycle. Cambridge University Press, Cambridge.
Mitchison, J, M. (1974). Sequences, pathways and timers. In: Cell Cycle Controls (Padilla, G.M., Cameron, I.L., & Zimmerman, A.M., Eds.), pp. 125–142. Academic Press, New York.
Mitchison, J.M. (1977). Enzyme synthesis during the cell cycle. In: Cell Differentiation in Microorganisms, Plant and Animals (Nover, L. & Mothes, K., Eds.), pp. 377–401. VEB Gustav Fischer Verlag, Jena.
Mitchison, J.M. (1988). Synchronous cultures and age fractionation. In: Yeast: A Practical Approach (Campbell, I. & Duffus, J.H., Eds.), pp. 51–63. IRL Press, Oxford.
Mitchison, J.M. (1989). Cell cycle growth and periodicities. In: Molecular Biology of the Fission Yeast (Nasim, A., Young, P., & Johnson, B.F., Eds.), pp. 205–242. Academic Press, San Diego.
Mitchison, J.M. & Nurse, P. (1985). Growth in cell length in the fission yeast *Schizosaccharomyces pombe*. J. Cell Sci. 75, 357–376.
Mitchison, J.M. & Vincent, W.S. (1965). Preparation of synchronous cultures by sedimentation. Nature, Lond. 205–989.
Mitchison, J.M., Creanor, J., & Novak, B. (1991). Coordination of growth and division during the cell cycle of fission yeast. Cold Spring Harbor Symp. Quant. Biol. 56, 557–565.
Mitchison, T.J. (1988). Microtubule dynamics and kinetochore function in mitosis. Annu. Rev. Cell Biol. 4, 527–549.
Molloy, G.R. & Schmidt, R.R. (1970). Studies on the regulation of ribulose-1,5-diphosphate carboxylase synthesis during the cell cycle of the eucaryote *Chlorella*. Biochem. Biophys. Commun. 40, 1125–1133.
Moreno, S., Nurse, P., & Russell, P. (1990). Regulation of mitosis by cyclic accumulation of $p80^{cdc25}$ mitotic inducer in fission yeast. Nature, Lond. 344, 549–552.
Murray, A.W. (1989). Cyclin synthesis and degradation during the embryonic cell cycle. J. Cell Sci. Suppl. 12, 65–76.
Murray, A.W. & Hunt, T. (1993). The Cell Cycle. An Introduction. W.H. Freeman, New York.
Nasmyth, K.A. (1993). Control of the yeast cell cycle by the Cdc28 protein kinase. Curr. Opin. Cell Biol. 5, 166–179.
Nasmyth, K.A. (1994). An egg-ocentric view of the cell cycle. Cell 78, 11–13.
Niklas, R.B. (1971). Mitosis. Adv. Cell Biol. 2, 225–297.
Niklas, R.B. (1988). The forces that move chromosomes in mitosis. Annu. Rev. Biophys. Chem. 17, 431–449.
Novak, B. & Mitchison, J.M. (1990). CO_2 production after induction synchrony of the fission yeast *Schizosaccharomyce pombe*: the origin and nature of entrainment. J. Cell Sci. 96, 79–91.
Nurse, P. (1990). Universal control mechanism regulatin onset of mitosis. Nature, Lond. 344, 503–508.
Nurse, P. & Fantes, P.A. (1981). Cell cycle controls in fission yeast: a genetic analysis. In: The Cell Cycle (John, P.C.L., Ed.), pp. 85–98. Cambridge University Press, Cambridge.
Nurse, P., Thuriaux, P., & Nasmyth, K.A. (1976). Genetic control of the cell division cycle in the fission yeast *Schizosaccharomyces pombe*. Molec. Gen. Genet. 146, 167–178.
O'Farrell, P.H. (1975). High resolution two-dimensional electrophoresis of proteins. J. Biol. Chem. 250, 4007–4021.
Prescott, D.M. (1955). Relations between cell growth and cell division. I. Reduced weight, cell volume, protein content and nuclear volume of *Amoeba proteus* from division to division. Exp. Cell Res. 9, 328–337.
Prescott, D.M. (1976). Reproduction of Eukaryotic Cells. Academic Press, New York.
Pringle, J.R. & Hartwell, L.H. (1982). The *Saccharomyces cerevisiae* cell cycle. In: The Molecular Biology of the Yeast *Saccharomyces*. Life Cycle and Inheritance (Strathern, J.N., Jones, E.W., & Broach, J.R., Eds.), pp. 97–142. Cold Spring Harbor Laboratory, Cold Spring Harbor.
Rao, P.N. & Sunkara, P.S. (1978). Cell fusion and regulation of DNA synthesis. In: Cell Cycle Regulation (Jeter, J.R., Cameron, I.L., & Padilla, G.M., Eds.), pp. 133–147. Academic Press, New York.

Rapkine, L. (1931). Sur les processes chimiques au cours de la division cellulaire. Annls. Physiol. Physicochim. Biol. 7, 382–418.

Rappaport, R. (1986). Establishment of the mechanism of cytokinesis in animal cells. Internat. Rev. Cytol. 105, 245–281.

Sachsenmaier, W. (1981). The mitotic cycle in *Physarum*. In: The Cell Cycle (John, P.C.L., Ed.), pp. 139–160. Cambridge University Press, Cambridge.

Satterwhite, L.L. & Pollard, T.D. (1992). Cytokinesis. Curr. Opin. Cell Biol. 4, 43–52.

Scherbaum, O. & Zeuthen, E. (1954). Induction of synchronous cell division in mass cultures of *Tetrahymena pyriformis*. Exp. Cell Res. 6, 221–227.

Schimke, R.T., Kung, A.L., Rush, D.F., & Sherwood, S.W. (1991). Differences in mitotic control among mammalian cells. Cold Spring Harbor Symp. Quant. Biol. 56, 417–425.

Shinagawa, A. (1983). The interval of the cytoplasmic cycle observed in non-nucleate egg fragments is larger than that of the cleavage cycle in normal eggs of *Xenopus laevis*. J. Cell Sci. 64, 147–162.

Shymko, R.M., Klevecz, R.R., & Kauffman, S.A. (1984). The cell cycle as an oscillatory system. In: Cell Cycle Clocks (Edmunds, L.N., Ed.), pp. 273–291. Marcel Dekker, New York.

Simchen, G. (1978). Cell cycle mutants. Annu. Rev. Genet. 12, 161–191.

Swann, M.M. & Mitchison, J.M. (1958). The mechanism of cleavage in animal cells. Biol. Rev. 33, 103–135.

Sweeney, B.M. & Hastings, J.W. (1958). Rhythmic cell division in populations of *Gonyaulax polyedra*. J. Protozool. 5, 217–214.

Tamiya, H., Iwamura, T., Shibata, K., Hase, E., & Nihei, T. (1953). Correlation between photosynthesis and light-independent metabolism in growth of *Chlorella*. Biochim. Biophys. Acta 12, 23–40.

Taylor, E.W. (1975). Some comments on the mechanism of mitosis. In: Molecules and Cell Movement (Inoue, S. & Stephens, R.E., Eds.), pp. 1–2. Raven Press, New York.

Terasima, T. & Tolmach, L.J. (1961). Changes in X-ray sensitivity of HeLa cells during the division cycle. Nature, Lond. 190, 1210–1211.

Tyson, J.J. (1979). Periodic enzyme synthesis: reconsideration of the theory of oscillatory repression. J. Theoret. Biol. 80, 27–38.

Tyson, J.J. (1983). Periodic enzyme synthesis and oscillator repression: why is the period of oscillation close to the cell cycle time? J. Theoret. Biol. 103, 313–328.

Tyson, J.J. & Sachsenmaier, W. (1984). The control of nuclear division in *Physarum polycephalum*. In: Cell Cycle Clocks (Edmunds, L.N., Ed.), pp. 253–272. Marcel Dekker, New York.

Veinot-Drebot, L.M., Johnston, G.C., & Singer, R.A. (1991). A cyclin protein modulates mitosis in the budding yeast *Saccharomyces cerevisiae*. Curr. Genet. 19, 15–19.

White, J.H.M., Barker, D.G., Nurse, P., & Johnston, L.H. (1986). Periodic transcription as a means of regulating gene expression during the cell cycle: contrasting modes of expression of DNA ligase genes in budding and fission yeast. EMBO J. 5, 1705–1709.

White-Cooper, H. & Glover, D.M. (1995). Regulation of the cell cycle during *Drosophila* development. In: Frontiers in Molecular Biology: Cell Cycle Control (Hutchison, C.J. & Glover, D., Eds.), pp. 264–296. IRL Press, Oxford.

Wick, M., Burger, C., Brusselbach, S., Lucibello, F.C., & Muller, R. (1994). Identification of serum-inducible genes; different patterns of gene regulation during $G_0 \rightarrow S$ and $G_1 \rightarrow S$ progression. J. Cell Sci. 107, 227–239.

Williams, N.E. & Macey, M.G. (1991). Is cyclin Zeuthen's "division protein?" Exp. Cell Res. 197, 137–139.

Williamson, D.H. & Scopes, A.W. (1961). Synchronisation of division in cultures of *Saccharomyces cerevisiae* by control of the environment. Symp. Soc. Gen. Microbiol. 11, 217–242.

Yashphe, J. & Halvorson, H.O. (1976). β-D-galactosidase activity in single cells during the cell cycle of *Saccharomyces lactis*. Science 191, 1283–1284.

Yeoman, M.M., Ed. (1976). Cell Division in Higher Plants. Academic Press, London.

Appendix 1

QUANTUM LEAPS

1869 Nuclei isolated.
1909 Maternal inheritance chlorophyll deficient areas in leaves of *Mirabilis jalapa.*
1928 Transformation demonstrated in mice.
1930s Tetranucleotide structure for DNA.
1934 Protein synthesis continued in enucleolate *Acetobularia.*
1938–1943 Basophilia in cytoplasm ascribed to RNA.
1938–1941 RNA component isolated from cytoplasm.
1941 Experiments with *Neurospora* suggest "ONE GENE, ONE ENZYME."
Postulation that protein synthesis is the reversal of protein synthesis.
ATP the energy source for protein synthesis?
1941–1942 Speculation that RNA is involved in protein synthesis.
1942 Protein turnover demonstrated *in vivo.* Macromolecular complexes containing RNA
isolated from cytoplasm. EM pictures.
1943 Particulate RNA-cytoplasmic fraction called "microsomes".
1944 TRANSFORMING PRINCIPLE SHOWN TO BE DNA.
1945 Name "Endoplasmic Reticulum" (ER) given to membranous basophilic cytoplasmic component seen in EM.
1948 Nuclear Magnetic Resonance first employed.
Oxidative energy needed for protein synthesis.
1948–1950 Amount DNA/cell constant for given species. [DNA] in diploid nuclei X2 that in haploid.
Early work on incorporation radioactive amino acids into proteins *in vivo* and *in vitro.*
Ratio A:T and G:C in DNA = 1.
1949 "Petite" colonies *S. cerevisiae* studied.
1950–1951 Amino acid uptake *in vivo* predominantly in microsome fraction.
1952 Only DNA transfered to *E. coli* from T_{even} phages during infection.
Amino acid incorporation *in vitro* predominantly into microsomal proteins.

1953 DOUBLE HELICAL STRUCTURE FOR DNA.
Amino acid sequence insulin completed.
Protein synthesis more specific biochemically than protein hydrolysis.

1954 First speculations on DNA coding published.
Uniparental inheritance antibiotic resistance.

1955 Small particulate components on ER seen in EM.
Polynucleotide phosphorylase isolated.
First evidence for mRNA in phage-infected *E. coli*.

1956 Microsome fraction due to fragmentation of ER and particles.

1956–1957 Two-subunit complexes isolated from ER by analytical centrifugation.
First NMR protein spectrum.

1957 Semiconservative replication shown in *Vicia faba* by autoradiography.
Amino acids activated for protein synthesis as amino-acyl adenylates.
No peptide intermediates involved in protein synthesis—which needs GTP.

1958 Semiconservative replication shown in *E. coli* by ^{15}N-labeling and ultracentrifugation.
DNA polymerase (DNA pol I) isolated.
Adaptor function for low MW RNA?
Amino-acyl adenylates transfered to CCA end tRNA.

1958–1959 Isolated ribosomes incorporated amino acids into proteins.
First X-ray structure of myoglobin.

1959 Chemical synthesis of defined polyribonucleotides.

1960 Double-stranded DNA/RNA hybrids constructed.
RNA synthesized on DNA template in *E. coli.*
First animal RNA polymerase isolated.
First indication secretory proteins synthesized on membrane-bound ribosomes.

1960–1961 Proof that mRNA necessary for protein synthesis.
Protein synthesis begins at N-terminal end peptide chain.

1961 Poly U codes for polyphenylalanine.
Both ribosomal subunits contain many different proteins but only one species of RNA.

1960–1962 DNA codons suggested from polynucleotide programming.

1961–1963 mRNA bound to small ribosomal subunit.

1962 Specificity of transfer of amino acids resides in tRNAs not the amino acid.
EM studies indicated presence DNA fibrils in plastids of *Chlamydomonas reinhardtii.*

1962–1963 Existence of polysomes carrying nascent proteins.

1962–1964	Biogenesis of ribosomes: nascent rRNA proteins verified.
1962	Net synthesis RNA demonstrated.
1963	Replication forks seen in *E. coli* by autoradiography.
	tRNA and nascent protein bound to large ribosomal subunit.
1963–1966	PRESENCE OF DNA IN PLASTIDS AND MITOCHONDRIA CONFIRMED.
1964	Triplet binding assay. CODONS FOR AMINO ACIDS FINALLY ESTABLISHED.
	Nucleolus the site of ribosomal RNA synthesis.
	Ribosomes can be dissociated into "core" particles and proteins.
	Met-tRNA and f met-tRNA involved in initiation protein synthesis in bacteria.
1964–1966	RNA pol I,II and III isolated from eukaryotes.
1965–1969	Stepwise degradation ribosomes and their subsequent reconstitution into active particles.
1966–1969	All secretory proteins synthesized on membrane-bound ribosomes.
1966	ALANYL tRNA SEQUENCED.
1966	"Wobble" hypothesis governing tRNA/mRNA interactions.
	Okazaki fragments detected during DNA replication.
1967–1969	Ribosomes formed in nucleus from proteins synthesized in cytoplasm and transported into nucleus.
1967–1971	Elongation steps in protein synthesis elucidated.
1968	5sRNA sequenced (120 nucleotides).
	Subunit exchange demonstrated in ribosomes.
1969	Cell division (*cdc*) mutants found in budding yeast.
	Ribosomal genes visualized in nucleolus.
	DNA pol I not primarily involved in DNA replication.
	Repetitive genes for rRNA and tRNA detected by hybridization.
	RNA primer required for DNA synthesis.
1969–1971	Termination steps in protein synthesis elucidated.
1969–1974	Initiation steps in protein synthesis elaborated.
1969	RESTRICTION ENZYMES ISOLATED.
1970	REVERSE TRANSCRIPTASE DISCOVERED.
1970s	Plastids and mitochondria only semiautonomous.
1970–1972	First recombinant DNA fragments from different sources.
	3D structure for insulin.
1972	First indications for "signal" peptide at N-terminus nascent protein.
1973	Restriction map of SV 40 presented.
	Calcium phosphate used to transfer DNA to mammalian cells.
	Protein domains recognized.
1974–1976	Location individual proteins on ribosomes by EM.
1974–1975	Binding sites for mRNA on rRNA identified.

1974 λ-phage introduced as vector for gene cloning.
1975 Southern blots devised.
1975–1977 Southern blotting technique adapted for transfer bacterial colonies and phage plaques.
1975–1979 SIGNAL HYPOTHESIS formulated and broadened.
1977 RNA splicing discovered. Introns found in eukaryotic nuclear genes.
pBR322 developed as cloning vector.
Cosmids introduced for cloning.
1978 Globin genes cloned.
1977–1980 TECHNIQUES FOR DNA SEQUENCING.
DNA introduced directly into cell nuclei.
1980s Introns found in yeast mitochondrial genome.
1981 Peptidyl transferase a multicomponent complex in ribosome.
Complete sequence mouse and human mitochondrial DNA (c.16.5kbp).
GENETIC CODE NOT UNIVERSAL.
1982 M13 introduced as cloning vector.
Electroporation used to transfer DNA.
Transgenic mice obtained expressing multiple copies rat growth hormone genes.
DNA transfer detected between plastids and mitochondria in maize.
First edition "Molecular Cloning," a Laboratory Manual.
1981–1983 "Signal" peptides characterized.
Mitochondrial and chloroplast proteins synthesized in organelles and cytosol.
1983 Modified retroviruses used for cloning.
Plastid DNA sequences detected in nuclear DNA.
1984 Pulsed-field gel electrophoresis introduced to separate large DNA fragments.
1985 Oligosaccharide sequencing developed.
1986 Polymerase chain reaction introduced.
1985–1987 Glycosylation changes established in disease.
1987 Plant cells transfected by microbombardment.
1988 *Taq* polymerase used for PCR.
1980s Yeast artificial chromosomes (YACs) constructed.
1989 Melon mitochondrial genome 2500 kbp.
1990 RNA editing found in mitochondria of trypanosomes.
1995 FIRST COMPLETE GENOME SEQUENCED FROM FREE-LIVING ORGANISM–*H. influenzae*.

Appendix 2

THE DNA CODE

	U		C		A		G	
U	UUU	Phe	UCU	Ser	UAU	Tyr	UGU	Cys
	UUC	Phe	UCC	Ser	UAC	Tyr	UGC	Cys
	UUA	Leu	UCA	Ser	UAA	Stop	UGA	Stop
	UUG	Leu	UCG	Ser	UAG	Stop	UGG	Trp
C	CUU	Leu	CCU	Pro	CAU	His	CGU	Arg
	CUC	Leu	CCC	Pro	CAC	His	CGC	Arg
	CUA	Leu	CCA	Pro	CAA	Gln	CGA	Arg
	CUG	Leu	CCG	Pro	CAG	Gln	CGG	Arg
A	AUU	Ile	ACU	Thr	AAU	Asn	AGU	Ser
	AUC	Ile	ACC	Thr	AAC	Asn	AGC	Ser
	AUA	Ile	ACA	Thr	AAA	Lys	AGA	Arg
	AUG	Met	ACG	Thr	AAG	Lys	AGG	Arg
G	GUU	Val	GCU	Ala	GAU	Asp	GGU	Gly
	GUC	Val	GCC	Ala	GAC	Asp	GGC	Gly
	GUA	Val	GCA	Ala	GAA	Glu	GGA	Gly
	GUG	Val	GCG	Ala	GAG	Glu	GGG	Gly

AUTHOR INDEX

SUBJECT INDEX

www.ingramcontent.com/pod-product-compliance
Ingram Content Group UK Ltd.
Pitfield, Milton Keynes, MK11 3LW, UK
UKHW020802290926
15470UKWH00003B/904

9 780762 300778